Advanced Digital Design Techniques

Vaibhav Taraate

Advanced Digital Design Techniques

High Speed and Low Power Design Techniques

Springer

Vaibbhav Taraate
Bengaluru, Karnataka, India

ISBN 978-981-95-4509-4 ISBN 978-981-95-4510-0 (eBook)
https://doi.org/10.1007/978-981-95-4510-0

This Springer imprint is published by the registered company Springer Nature Singapore Pte Ltd.
The registered company address is: 152 Beach Road, #21-01/04 Gateway East, Singapore 189721, Singapore

Dedicated to
Supreme Lord of Intelligence
and
Proxelera Team

Preface

The advanced digital design techniques and the optimization for the area, speed and power are discussed in this book and useful to the engineers during the SoC and IP designs.

The book has 13 chapters and useful to understand the constraints such as area, speed, and power and optimization of the design to have the lesser area, lesser power, and max speed. There is always trade-off between the area and power.

The low power, multiple clock, and power domain details are discussed with the practical scenario and useful to the engineers, students, and professionals.

Chapter 1 **Introduction to Digital Circuits:** This chapter is focused on the basic fundamentals of the digital circuit and the basic logic gates and use of them in the design. The chapter is also useful to understand the basics of the area, speed, and power constraints and practical scenarios.

Chapter 2 **Area Optimized Designs:** The chapter is useful to understand the resource utilization, performance improvement by limiting the area. The chapter covers the area optimized designs, clock mux, scan mux, and optimization concepts with the practical scenarios.

Chapter 3 **The Data and Control Paths:** The chapter is useful to understand the data and control paths in the design and how we can use the area optimization techniques while passing data from the data path for the various control signals. The chapter also covers the details of the optimization for the data and control path and use of the universal logic elements which is multiplexers and design using multiplexers.

Chapter 4 **The Area Optimization: Case Study:** This chapter is useful to understand the area optimization and logic design using minimum number of logic elements. For any kind of the ASIC design the least area indicates the lesser number of logic gates to implement the design functionality. By considering this as an objective the design optimization scenarios are included in this chapter.

Chapter 5 **Practical Scenarios: Area Optimization:** The chapter covers the practical scenarios, area optimization techniques, logic duplications, cascade versus parallel logic, delay optimization, and area optimized designs.

Chapter 6 **Sequential Design:** The sequential design using flip-flops is discussed in this chapter and basics discussed are useful during the speed optimization discussion in the subsequent chapters. This chapter mainly focuses on the basics of the sequential design and how to design the optimized sequential circuits.

Chapter 7 **Timing Parameters and Optimization Scenarios:** This chapter is useful to understand the timing parameters, insertion delays, timing paths, and the slack calculations. The chapter is also useful to understand the setup and hold slack calculations and various design scenarios and how to fix the issues. The chapter is useful to understand the basics of multiple clock domain designs, glitch free clock mux design, and the skews in the design.

Chapter 8 **Timing Analysis and Speed Optimization:** This chapter is useful to understand the timing details, timing analysis, and the speed optimization, improvement techniques. The chapter covers the practical scenarios for the register balancing, pipelining, and the speed improvement of the design.

Chapter 9 **Synchronizers and High Speed Designs:** The various synchronizers which are used in the multiple clock domains are discussed in this chapter. The chapter covers the control path, data path synchronizers, and practical considerations and scenarios during the design. The chapter also covers the reset synchronizers and the strategies for the deassertion and reset domain crossing issues.

Chapter 10 **FSM and FSMD:** The chapter covers the FSM and FSM with the data path and various techniques for the efficient FSM designs. The chapter also covers the glitch free and optimized FSM designs and practical considerations while design of the FSMD.

Chapter 11 **Low Power Design and Optimization Techniques:** The chapter covers the low power design techniques and practical scenarios. The chapter also covers the power optimization, multiple power domain designs, and the design considerations.

Chapter 12 **DFT-Friendly Designs and Practical Scenarios:** The chapter is useful to understand the system design practical scenarios and system design consideration. The chapter also covers the details of the area, speed, and timing constraints during the system design. Chapter is useful during the design of the digital system which has IPs, SoC components. Also, the main focus of this chapter is to have detail understanding of the DFT friendly designs and test circuits used in the design and their optimization to have higher performance of the digital system.

Chapter 13 **System Design Case Study and Guidelines:** The chapter covers the application of the Chaps. 1–12 in the system design. Chapter is useful during the design of the digital system which has Ips, SoC components. The design guidelines are discussed in this chapter.

The book covers the various practical scenarios and design considerations which are useful during the SoC and IP designs.

This book is useful to the postgraduate engineers, logic design engineers, VLSI professionals, SoC, and IP designers.

Bengaluru, India Vaibhav Taraate

Acknowledgements

Over the past decade, we have witnessed the rapid growth of the design techniques, and there have been many evolutions in the chip design. By considering the latest evolution, I thought to document the advanced digital design techniques in this manuscript.

I am thankful to all those entrepreneurs, design/verification engineers, and managers with whom I worked in the past, almost around 25 years.

I am thankful to my Proxelera team for the direct and indirect support. Especially thankful to my friends, well-wishers, and my family members. Special thanks to Ravi, Mohan, Davinder, and Shailesh for their best wishes and for their valuable help during the work.

Special thanks to Somi, Siddhesh, and Kajal for their faith and belief. Especially thankful to my dearest friends Neeraj, Deepesh, Annu, and Tannu for the motivation and great support.

Finally, I am thankful to the Springer Nature staff, especially Swati Meherishi, Divya Meiyazhagan, and Ashok Kumar, for their great support during the various phases of the manuscript.

Special thanks in advance to all the readers, engineers, and professionals!

Contents

1 Introduction to Digital Circuits 1
1.1 The Practical Systems 1
1.2 Basics of Number Representation 2
1.2.1 Digital Logic 3
1.2.2 Area for the Design 4
1.2.3 Speed of the Design 5
1.2.4 Power 6
1.3 Concept of the Delay 8
1.4 Logic Gates 9
1.5 Concept of Delays in the Digital Circuits 14
1.6 Basics of thc Dclay Optimization 15
1.7 Universal Logic Gates 15
1.8 Important Highlights 16
1.9 Puzzle 16

2 Area Optimized Designs 17
2.1 Design Using NAND Gate 17
2.2 The Multiplexer as a Universal Logic 19
2.3 The 2:1 Mux Using the Minimum Number of NAND Gates 20
2.4 Design of the n-Bit Adder 22
2.5 Design of the n-Bit Subtractor 22
2.6 Design of the n-Bit Adder-Subtractor 24
2.7 Area Optimized Design 24
2.8 Design Optimization and Resource Sharing 28
2.9 Practical Scenarios: CLK MUX 28
2.10 Mux-Based Chain for the Selection 30
2.11 The Scan Mux 31
2.12 Important Highlights 31
2.13 Puzzle 33

3 The Data and Control Paths 35
3.1 Data Path 35
3.2 Control Path 37
3.3 Multiplexer as a Control Element 37
3.3.1 Complement Logic 39
3.3.2 AND Logic 39
3.3.3 NAND Logic 40
3.3.4 OR Logic 41
3.3.5 NOR Logic 42
3.3.6 XOR Logic 43
3.3.7 XNOR Logic 44
3.4 Universal Logic Gate 45
3.4.1 The Custom Universal Gate and Use in Optimization 46
3.4.2 The Custom Universal Gate-II and Use in Optimization 46
3.5 The Optimization Using Mux 48
3.5.1 The 4:1 Mux as a Data Path Selector 48
3.5.2 The Optimized Design of 4:1 Mux 49
3.6 Important Highlights 52
3.7 Puzzle 52

4 The Area Optimization: Case Study 53
4.1 Resource Sharing 53
4.2 Arithmetic Unit and Area Optimization Scenarios 56
4.2.1 Design Optimization Using Least Resources 58
4.3 The Logic Unit Design 60
4.4 The ALU Design 63
4.5 The Efficient ALU Design: Area Optimization 65
4.6 Important Takeaways 68
4.7 Puzzle 68

5 Practical Scenarios: Area Optimization 69
5.1 Cascade Mux 4:1 69
5.2 Propagation Delay 70
5.2.1 Input Swapping: Late Versus Faster Arrival Signals 71
5.3 Hazards 71
5.4 Glitches 75
5.5 Decoding of the Data Input 76
5.6 Encoding of the Data Input 81
5.7 Logic Duplication 84
5.8 Grouping 85
5.9 Practical Scenarios 86
5.10 Important Highlights 87
5.11 Puzzle 90

6 Sequential Design ... 91
6.1 Level Sensitive Latches ... 91
6.1.1 Positive Level Sensitive Latch ... 91
6.1.2 Negative Level Sensitive D Latch ... 92
6.2 D Flip Flops: Edge Sensitive Elements ... 93
6.2.1 Positive Edge Sensitive D Flip-Flop ... 93
6.2.2 Negative Edge Sensitive D Flip-Flop ... 95
6.3 Frequency Divider ... 96
6.4 Counters ... 97
6.4.1 Synchronous Binary up Counter ... 98
6.4.2 Synchronous Binary Down Counter ... 100
6.4.3 Gray Counters ... 102
6.5 Ring Counter ... 105
6.6 Twisted Ring Counter ... 108
6.7 Important Highlights ... 111

7 Timing Parameters and Optimization Scenarios ... 113
7.1 D Flip-Flop Timing Parameters ... 113
7.2 Input Insertion Delays ... 114
7.3 Output Insertion Delays ... 115
7.4 Reg to Reg Path ... 116
7.5 Setup Slack ... 118
7.5.1 Positive Setup Slack ... 118
7.5.2 Negative Setup Slack ... 119
7.6 Hold Violations ... 119
7.7 Asynchronous Circuits ... 119
7.8 Practical Scenarios ... 120
7.9 Clock Skew ... 120
7.10 Can We Control the Clock Polarity? ... 120
7.11 Clock Domains in the Design ... 121
7.12 Clock Mux ... 124
7.13 Level Synchronizer to Avoid the Glitches ... 127
7.14 Important Highlights ... 127

8 Timing Analysis and Speed Optimization ... 129
8.1 Clock Skew ... 129
8.2 Frequency Calculations ... 129
8.2.1 Design I ... 131
8.2.2 Design II ... 131
8.2.3 Design III ... 133
8.3 Skew in the Design and Impact on the Frequency ... 133
8.4 Timing Optimization ... 135
8.5 Pipelining ... 137
8.6 Important Highlights ... 138
8.7 Puzzle ... 138

9 Synchronizers and High Speed Designs 139
9.1 Level Synchronizers 139
9.2 Multi-bit Signal Passing 139
9.3 Handshake Synchronizers 141
9.4 Mux Synchronizer 141
9.5 Pulse Synchronizer 143
9.6 Reset Synchronizers 143
9.6.1 Active Low Asynchronous Reset 144
9.6.2 Active High Asynchronous Reset 144
9.7 Data Path Synchronizers 145
9.7.1 Faster to Slower Clock Domain 145
9.7.2 Faster to Slower Clock Domain with Idle Cycles 146
9.7.3 Slower to Faster Clock Domains 147
9.7.4 Slower to Faster Clock Domain with Idle Cycle 148
9.7.5 Same Frequency in Both Clock Domain 149
9.8 Important Highlights 149
9.9 Puzzle 150

10 FSM and FSMD 151
10.1 FSM Designs 151
10.1.1 Moore FSM 151
10.1.2 Mealy FSM 152
10.1.3 Moore Versus Mealy FSM 153
10.2 State Encoding Methods 153
10.3 Moore FSM Design 154
10.4 Mealy FSM Design 158
10.5 Moore 101 Overlapping Sequence Detector Design 162
10.6 Mealy FSM Design 167
10.7 FSMD 171
10.8 Important Highlights 172
10.9 Puzzle 172

11 Low Power Design and Optimization Techniques 173
11.1 Introduction 173
11.2 Low Power Design Challenges 175
11.3 Clock Gating Cells 175
11.4 Glitch Free Clock Gating Cell 177
11.5 Low Power Architecture Design Considerations 177
11.5.1 Power Gating 177
11.5.2 Frequency Gating and Dynamic Frequency Scaling 178
11.5.3 Multivoltage (Multi-Vdd) Design 178
11.5.4 Dynamic Voltage and Frequency Scaling 178
11.6 The UPF 179
11.7 The Power Management Unit 179
11.8 Low Power Aware SoC Designs 181
11.9 Important Highlights 182

12 DFT-Friendly Designs and Practical Scenarios 185
12.1 Scan Flip-Flop 185
12.2 Scan Chains in the Design 186
12.3 Scan Chains for the Low Power Designs 188
12.4 The Power Management Circuits and Test Scenarios 189
12.4.1 The Clock Gating Test 190
12.4.2 Power Switch and Testing 190
12.4.3 Power Switch and Testing 191
12.5 Important Highlights 191

13 System Design Case Study and Guidelines 193
13.1 Pipelined Processor Micro-architecture 193
13.2 SDRAM Controller Micro-architecture Design 196
13.3 Low Power Design Considerations 199
13.4 DFT-Friendly Design Considerations 199
13.5 AI ML-Based Design 202
13.6 Important Highlights 203

Bibliography 205

Index 207

About the Author

Vaibbhav Taraate, the RTL Lead at Proxelera Pvt. Ltd., is a highly qualified professional with a B.E. (Electronics) degree from Shivaji University, Kolhapur (1995). Notably, he was awarded a Gold Medal for securing the top position across all engineering branches. Furthering his academic pursuits, Vaibbhav completed his M.Tech. in Aerospace Control and Guidance from the prestigious Indian Institute of Technology (IIT) Bombay, India, in 1999.

With a rich background spanning over 21 years, Vaibbhav specializes in semi-custom ASIC and FPGA design, with a focus on HDL languages such as Verilog, VHDL, and SystemVerilog. Throughout his career, he has collaborated with leading multinational corporations in various capacities, including consultant, senior design engineer, and technical manager.

Vaibbhav's areas of expertise encompass a wide range of domains, including RTL design utilizing Verilog and SystemVerilog, intricate FPGA-based design, low-power design strategies, synthesis and optimization techniques, static timing analysis, system design employing microprocessors, high-speed VLSI designs, and the architectural design of complex SOCs. His extensive experience and diverse skill set make him a valuable asset in the field of semiconductor design and engineering.

Chapter 1
Introduction to Digital Circuits

For high speed and low power ASIC designs, focus is on minimizing the delays, signal integrity, power integrity, through techniques like careful routing, impedance matching, and power delivery network optimization. By considering this we always plan the strategy to have the low power, high speed, and the DFT friendly designs. To design the efficient digital circuits, we need to understand the constraints and basic digital elements. This chapter is focused on the basic fundamentals of the digital circuit and the basic logic gates and use of them in the design.

1.1 The Practical Systems

If we consider the system design then the systems have the analog circuits and digital circuits. For example, consider the following system, here we have the analog input and Analog to Digital Converter (ADC) samples the input at the required sampling frequency. At the end of the conversion, it generates the interrupt to the processor.

The processor uses the digital data and it performs the required computations on the digital data and gives data to the Digital to Analog Converter (DAC).

DAC uses the valid data from the processor and converts it into the analog output (Fig. 1.1).

The analog signal is function of the time and is sampled by the ADC and the digital output from ADC which is in the form of 0,1 is processed by the processor. The processed data by the processor is used by the DAC and final output is an analog output from the system.

The scope of the manuscript is to have detailed discussion on the advanced digital design techniques and their use. For the analog design readers can read the analog design books.

V. Taraate, *Advanced Digital Design Techniques*,
https://doi.org/10.1007/978-981-95-4510-0_1

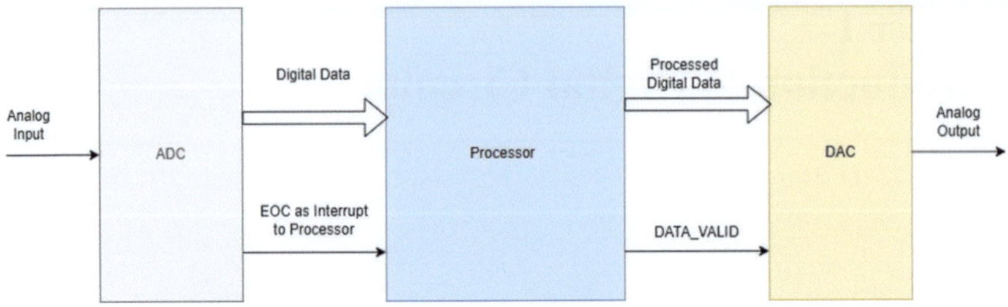

Fig. 1.1 System design components

By considering the scope of this manuscript the following sections of this chapter focus on the basics of the design elements as a quick reference. Let us start with the basics of binary, hex numbers, and the logic gates.

1.2 Basics of Number Representation

The objective of this manuscript is to discuss the advanced digital design techniques and the reader should have background of the basics of digital circuits and number systems. For easy reference refer Table 1.1.

Table 1.1 Decimal to binary number

Decimal number	B3	B2	B1	B0
0	0	0	0	0
1	0	0	0	1
2	0	0	1	0
3	0	0	1	1
4	0	1	0	0
5	0	1	0	1
6	0	1	1	0
7	0	1	1	1
8	1	0	0	0
9	1	0	0	1
10	1	0	1	0
11	1	0	1	1
12	1	1	0	0
13	1	1	0	1
14	1	1	1	0
15	1	1	1	1

Table 1.2 Hex and its binary equivalent

HEX	B3	B2	B1	B0
0	0	0	0	0
1	0	0	0	1
2	0	0	1	0
3	0	0	1	1
4	0	1	0	0
5	0	1	0	1
6	0	1	1	0
7	0	1	1	1
8	1	0	0	0
9	1	0	0	1
A	1	0	1	0
B	1	0	1	1
C	1	1	0	0
D	1	1	0	1
E	1	1	1	0
F	1	1	1	1

Hex numbers 0–9 and A to F, octal from 0 to 7, and binary coded decimal (BCD) from 0 to 9.

Hex number and its binary equivalent are shown in Table 1.2.

1.2.1 Digital Logic

We have two types of the designs: Combinational or combinatorial and sequential.

1. ***Combinational logic*****:** In the combinational logic an output is function of the present input. Following are examples of the combinational logic:

   a. **Logic gates.**
   b. **Arithmetic resources such as adders, subtractors, comparators.**
   c. **Multiplexers.**
   d. **Decoders.**
   e. **Demultiplexers.**
   f. **Encoders.**

2. **Sequential logic:** In the sequential design an output is function of the present input and past output. The examples of the sequential logic are:

   a. **Flip-flops.**
   b. **Counters.**

c. **Shift-registers.**
d. **Memory elements.**

Considerations during the design: During the design we consider the following:

1. Area optimized design.
2. Low power aware architecture and design.
3. High speed design.
4. DFT friendly design.

Most of the manuscript chapters focus on the use and application of the digital logic elements to have the design with the optimized area, low power, high speed, and DFT friendly designs.

In the ASIC design we have the constraints as area, speed, and power.

1.2.2 Area for the Design

The area of the design is one of the design constraints.

1.2.2.1 Area Constraints

- Area constraints are limitations or restrictions placed on the physical size of the design during the ASIC design process.
- They define the maximum area that the design can occupy on the silicon die.
- These constraints are used to ensure that the final design fits within the available space and meets performance requirements.
- They are especially important in applications where space is limited, such as processing, computing.

1.2.2.2 Importance of the Area Constraints

Following few of the details which we consider during the design.

1. **Resource Utilization**: Area constraints help designers optimize the use of available resources on the chip.
2. **Performance**: By limiting the area, designers can ensure that the design meets timing and performance requirements.
3. **Cost**: Minimizing area can lead to lower manufacturing costs, as smaller chips are generally cheaper to produce.
4. **Layout and Routing**: Area constraints guide the placement and routing of design elements, ensuring that the final layout is feasible.

1.2.2.3 Use of the Area Constraints in the Design

Now let us understand how we will use the area constraints in the design.

1. **Synthesis Tools**: Synthesis tools use area constraints to guide the mapping of the design to the target technology.
2. **Floorplanning**: Area constraints are used during floorplanning, a critical step in the physical design flow, where the overall layout of the chip is planned.
3. **Place and Route**: Place and route tools use area constraints to guide the placement of cells and the routing of interconnects.

1.2.2.4 Types of Area Constraints

Following are few of the area constraints:

1. **Maximum Area**: Specifies the maximum area that the design can occupy.
2. **Minimum Area**: Specifies the minimum area that the design should occupy.
3. **Outline Constraints**: Define the boundaries of the design.
4. **Placement Constraints**: Specify the location of specific blocks or cells.
5. **Routing Constraints**: Specify the routing paths for specific nets.

1.2.2.5 Techniques for Area Optimization

Few of the techniques for the area optimization are listed below

1. **EDA Tools**: Electronic Design Automation (EDA) tools, such as those from Synopsys, Cadence, and Mentor Graphics, provide features for specifying and managing area constraints.
2. **Synopsys Design Constraints (SDC)**: The Synopsys Design Constraints (SDC) format is used to specify timing and area constraints.
3. **Floorplanning**: Floorplanning tools allow designers to visualize and manipulate the layout of the chip, taking into account area constraints.
4. **Physical Synthesis**: Physical synthesis tools optimize the design for area and timing, taking into account area constraints.

1.2.3 Speed of the Design

The speed of the design is one of the important parameters and the speed is limited due to the inertial gate delay or due to cascading of the number of logic stages. As shown in Fig. 1.2, if we consider the CMOS NOT gate then at the output there is formation of the stray capacitance and the inertial delay of the NOT gate is due to the time required for charging and discharging of the capacitor. The inertial delay is the propagation delay and defined as the amount of time required for the output to get the valid logic level after change in the input.

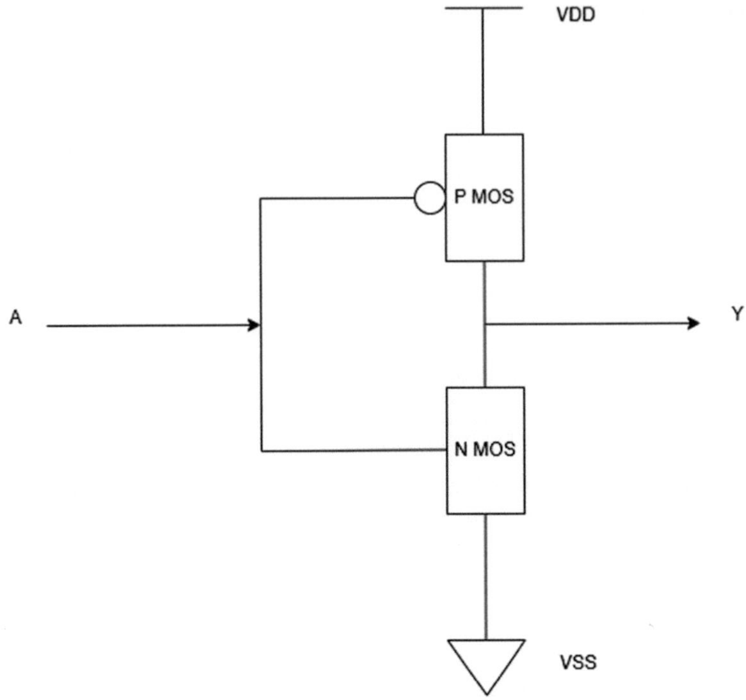

Fig. 1.2 CMOS NOT gate

For the sequential design, the speed is mainly depends upon the timing parameters of the sequential design elements such as setup time, hold time, clock to q delay (propagation delay of the flip-flop). Refer the subsequent chapters for more details about the timing parameters.

There are various speed improvement techniques used during the design stages and few important techniques which are discussed are:

1. Balancing the register timing.
2. Pipelining.
3. Register balancing.
4. Optimizing for the timing paths in the design.
5. Use of the parallelism if area is not important consideration.

1.2.4 Power

The power is an important design consideration. Most of we have come across the terms such as switching power, static, and dynamic power. The power dissipation should be minimum, and the book also focuses on the low power architecture design techniques and their role at various stages of the design. Consider Fig. 1.2, as shown as the stray capacitance is formed at the output the power is specified as

$$\boldsymbol{Power = \alpha Cs VDD^2 f}$$

where α = Switching factor

Cs = Load or stray capacitance

VDD = Supply voltage.

To have the minimum power the load capacitance should be minimum, and the voltage should be minimum but as we can't compromise on the speed of the design we need to balance between the desired power and speed of the design (Fig. 1.3).

There is always the tradeoff between the speed and power and the main goal of the logic design team is to have the balance act to achieve the desired speed and power for the design.

Few of the power optimization techniques are listed below

1. Use of low power cells in the design.

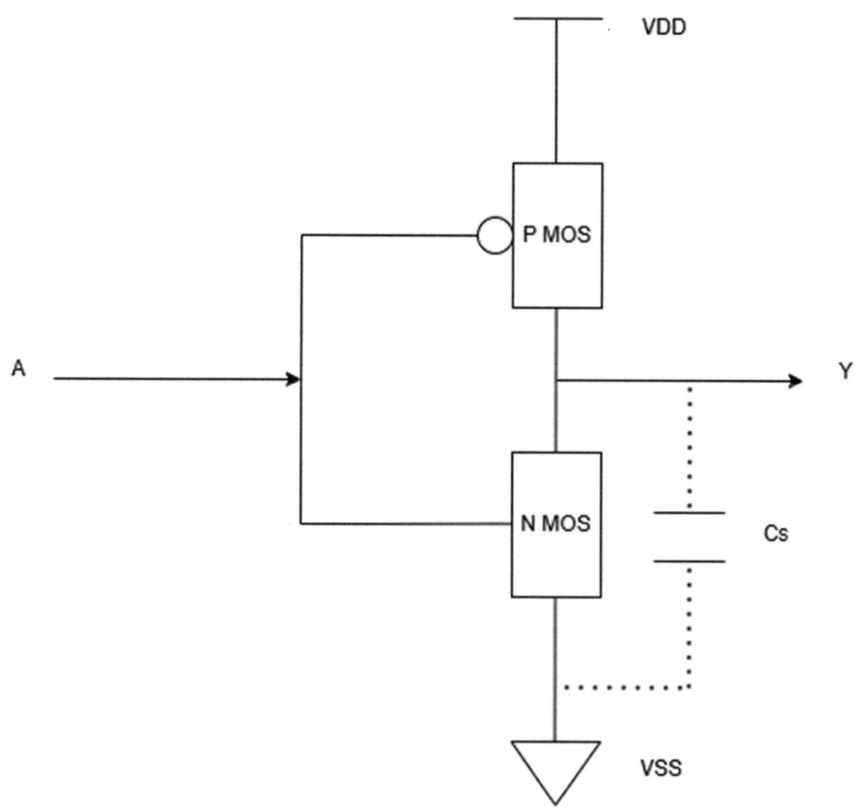

Fig. 1.3 Logic with the stray capacitance

2. Low power aware architecture design.
3. Clock gating for the sequential logic.
4. Voltage and frequency scaling.

These techniques are discussed in the subsequent chapters.

Now as we have understood the constraints, goals, and objectives during the logic design. As stated earlier the basic logic elements are logic gates and the following section discusses about them.

1.3 Concept of the Delay

The propagation delay is the amount of time for the logic gate to get the valid output after changes in the input.

Ideal System: If we consider the ideal waveform then the delay is the delay between the edges where the transition happens (Fig. 1.4).

In the practical scenario the propagation delay of the NOT gate is represented as a fall delay (Tfall) and rise delay (Trise). The fall and rise delays are shown in the waveform below (Fig. 1.5).

Practically both the rise and fall delays can have different value.

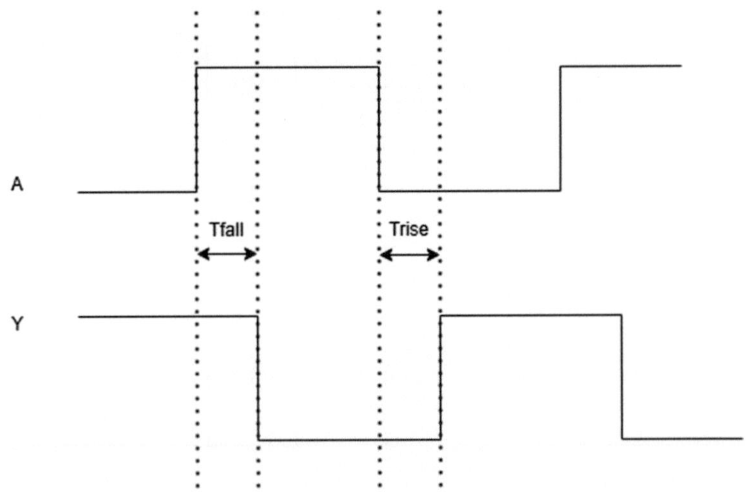

Fig. 1.4 Ideal waveform of NOT gate with fall and rise time

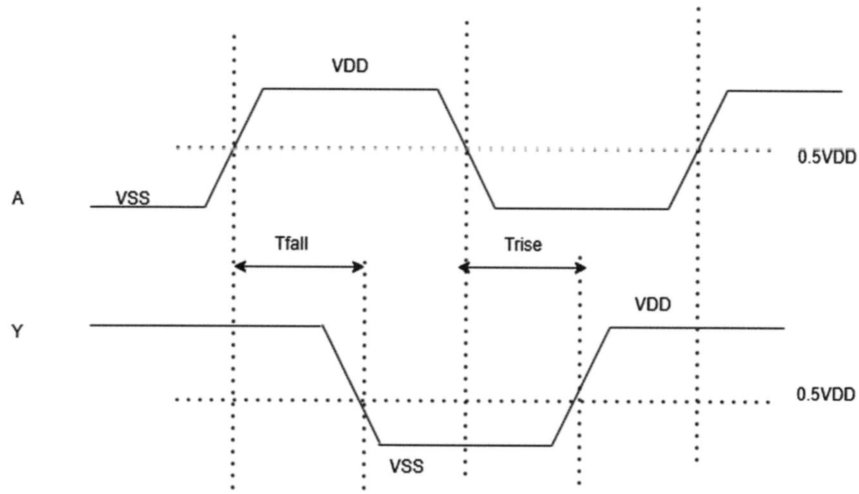

Fig. 1.5 Timing diagram of the rise and fall delay

1.4 Logic Gates

To build the efficient digital logic we need to have basic fundamental understanding of the logic gates. This section we will discuss the basic logic gates such as NOT, OR, NOR, AND, NAND, XOR, and XNOR.

1. **NOT gate**

NOT has single input and single output and complements the digital input. That is digital output is complement of the digital input. Output is 180-degree phase shifted of the input.

The truth-table of the NOT gate is shown in Table 1.3. It has input A and an output Y. The Boolean expression is given by

$$Y = \overline{A}.$$

The NOT gate is shown in Fig. 1.6 and the output is complement of an input. The

Table 1.3 Truth-table of NOT gate

A	Y
0	1
1	0

Fig. 1.6 NOT gate

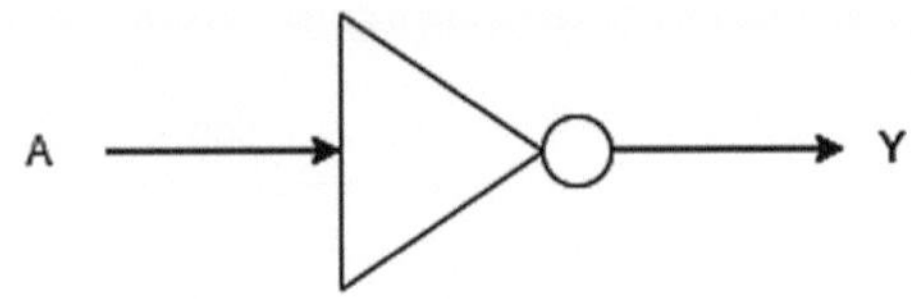

Table 1.4 Truth-table of OR gate

A	B	Y
0	0	0
0	1	1
1	0	1
1	1	1

complement of the logic 1 is 0 and complement of the logic 0 is logic 1.

2. **OR gate**

OR is the logical OR of the inputs. The two input OR gate performs logical OR of the two binary inputs. Output is A OR B.

The truth-table of the OR gate is shown in Table 1.4. It has inputs A, B and an output Y. The Boolean expression is given by

$$Y = A + B.$$

Figure 1.7 is symbol of OR gate and it indicates that either A or B should be logic 1 to get an output Y as logic 1.

3. **NOR gate**

The NOR is NOT of OR. If one of the inputs is logic 1 then an output of NOR gate is logic 0. The NOR gate is universal logic gate. By using the number of NOR gates any Boolean function can be implemented.

The truth-table of NOR gate is shown in Table 1.5 and has inputs as A, B and output as Y. The Boolean expression of the NOR gate is

$$Y = \overline{A + B}.$$

The NOT of OR is shown in Fig. 1.8 which is the cascade of OR and NOT. ***The issue is the larger propagation delay due to cascading of the OR and NOT. So,***

Fig. 1.7 OR gate

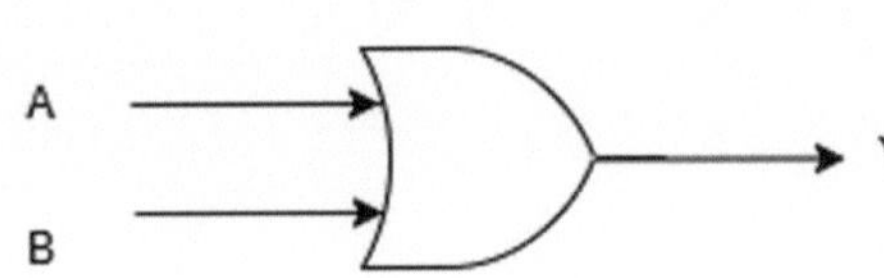

Table 1.5 Truth-table of NOR gate

A	B	Y
0	0	1
0	1	0
1	0	0
1	1	0

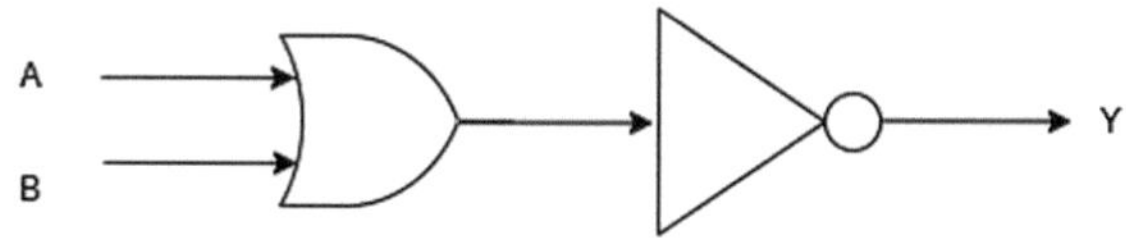

Fig. 1.8 NOT of OR

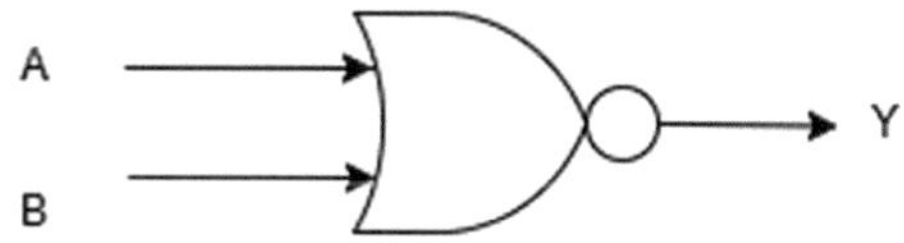

Fig. 1.9 NOR gate. *Note* Use the minimum number of the NOR gates to implement any combinational logic

during design avoid the cascade logic. If the delay of each gate is 0.25 ns then the propagation delay of the logic is 0.5 ns.

The symbol of the NOR gate is shown in Fig. 1.9.

4. **AND Gate**

The AND gate output is logic 1 when both the inputs A, B are at logic 1. The Boolean expression of 2-input AND gate is given by

$$Y = A.B.$$

The truth-table is shown in Table 1.6. As shown in the truth-table when both the inputs A, B are logic 1, an output of the AND gate is logic 1. If one of the inputs is logic 0, the output of AND gate remains at logic 0.

The symbol of the 2-input AND is shown in Fig. 1.10 and as shown the AND gate has two inputs A, B and an output as Y.

Table 1.6 Truth-table of AND gate

A	B	Y
0	0	0
0	1	0
1	0	0
1	1	1

Fig. 1.10 AND gate

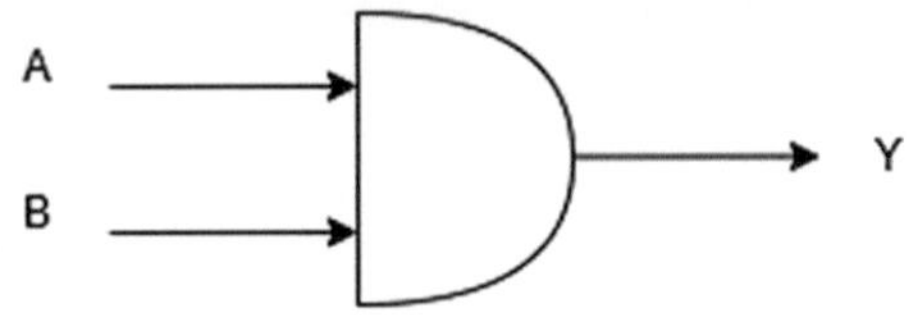

5. NAND Gate

NAND is NOT of AND. The truth-table of 2-input NAND gate is described in Table 1.7. The Boolean expression is given by

$$Y = \overline{A.B}.$$

If A = 1 and B = 1 then only an output of NAND gate is 0.

The NOT of AND is shown in Fig. 1.11. The cascade stages of logic will have larger delays. Here if each gate has delay of 0.25 ns then the overall propagation delay is 0.5 ns.

The NAND gate symbol is shown in Fig. 1.12 and as shown it has two inputs A, B and an output as Y.

6. XOR Gate

The Exclusive OR (XOR) truth-table is shown in Table 1.8. As shown if both the inputs are not equal then two input XOR gate output is logic 1. The Boolean expression of XOR gate is

Table 1.7 Truth-table of NAND gate

A	B	Y
0	0	1
0	1	1
1	0	1
1	1	0

Fig. 1.11 NOT of AND

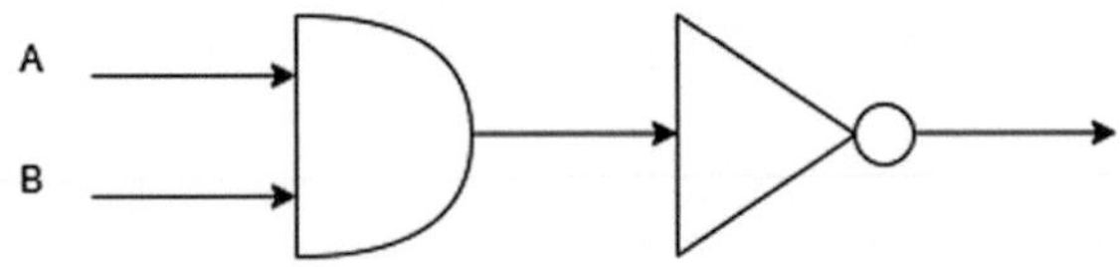

Fig. 1.12 NAND gate

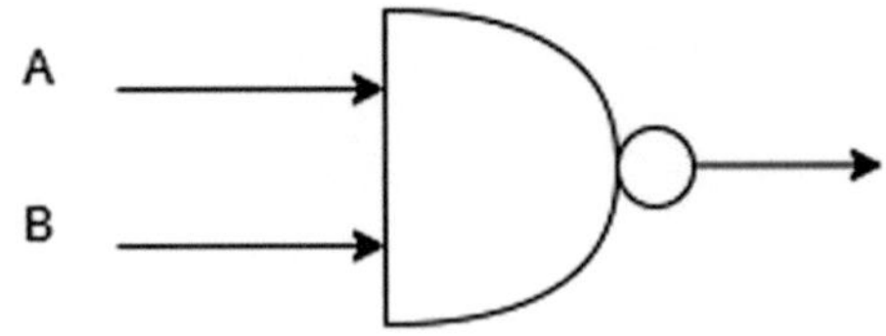

Table 1.8 Truth-table of XOR gate

A	B	Y
0	0	0
0	1	1
1	0	1
1	1	0

$$Y = A \oplus B.$$

The symbol of the XOR gate is shown in Fig. 1.13 and as shown the two input XOR gate has inputs A, B and an output as Y.

7. **XNOR Gate**

The XNOR is NOT of XOR. The XNOR is denoted by $\odot$ *(Pronounced as EXNOR)*. The logic expression is given by

$$Y = A \odot B.$$

The EXNOR or exclusive NOR or XNOR are few names we use during the design. The XNOR is NOT of XOR. The XNOR using the XOR, NOT is shown in Fig. 1.14.

The truth-table of the XNOR gate is shown in Table 1.9. As shown the output of two input XNOR gate is logic 1 when both the inputs are equal.

As discussed earlier avoid the cascading of the stages as it increases the propagation delay.

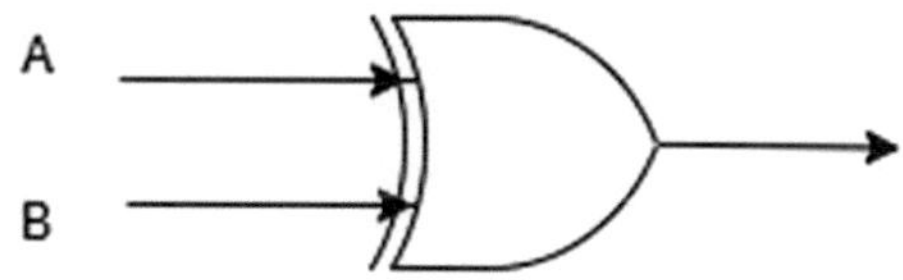

Fig. 1.13 XOR gate

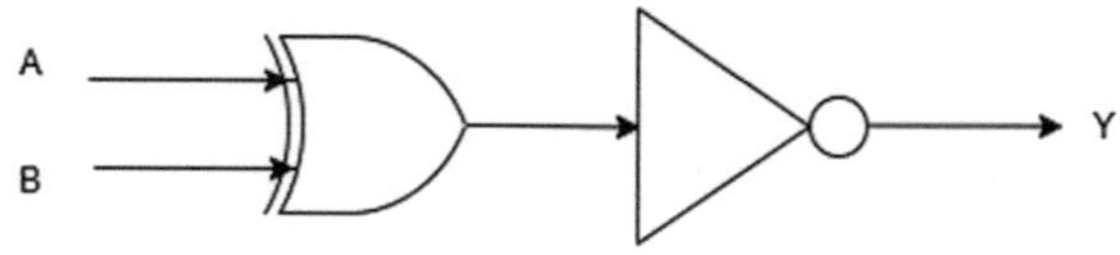

Fig. 1.14 NOT of XOR

Table 1.9 Truth-table of XNOR gate

A	B	Y
0	0	1
0	1	0
1	0	0
1	1	1

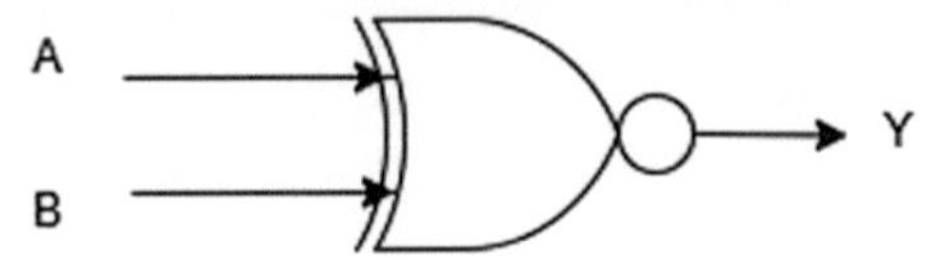

Fig. 1.15 XNOR gate

The symbol of the XNOR gate is shown in Fig. 1.15 and as shown the logic has A, B inputs and output as Y.

1.5 Concept of Delays in the Digital Circuits

The propagation delays in the digital circuit are defined as, it is minimum amount of the time required for the digital circuit to generate valid output after change in the input. For example, the propagation delay of the logic gate is 1 ns; it indicates that if input changes at time instance $t = t1$ then an output of logic gate changes at $t1 +$ 1 ns, where ns is nanosecond.

Now let us consider the practical scenario: The logic gates are cascaded as shown in the following Fig. 1.16. In such condition if the propagation delay (t_{pd}) is addition of each gate delay.

If the delay of each logic gate is t_{pd} then the overall propagation delay is $3 * t_{pd.}$ In Fig. 1.16 three AND gates are cascaded and the

$$
\begin{aligned}
Y1 &= A.B \\
Y2 &= Y1.C \\
Y2 &= A.B.C \\
Y &= Y2.D \\
Y &= A.B.C.D.
\end{aligned}
$$

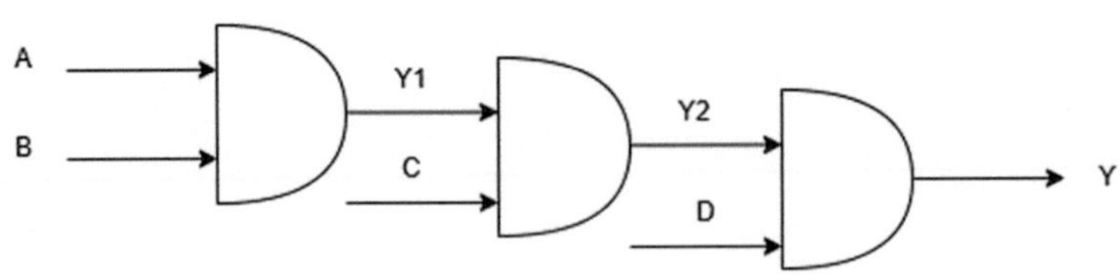

Fig. 1.16 Cascade logic

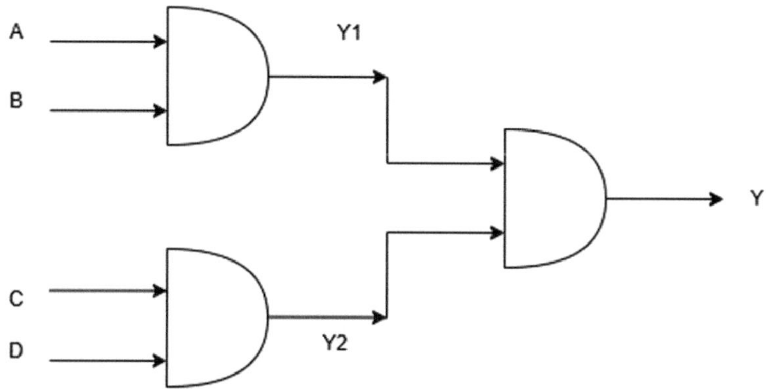

Fig. 1.17 Optimized four-input AND gate using 2-input AND gates

1.6 Basics of the Delay Optimization

As discussed, in the previous section the cascading results into the higher propagation delay. To optimize the delay in the digital circuit we can use the grouping technique.

$$Y1 = A.B$$
$$Y2 = C.D$$
$$Y = Y1.Y2$$
$$Y = (A.B).(C.D)$$

In Fig. 1.17 two parallel AND gates have delay of the tpd and the output gate has delay of tpd. So, overall delay is 2 * t_{pd}. If delay of each 2-input AND gate is 1 ns then the overall propagation delay is 2 ns. This is achieved by grouping of the terms to have the parallel logic.

1.7 Universal Logic Gates

The universal logic gates are NAND, NOR, and by using these gates we can design any combinational logic, and hence they are called as universal logic gates. Following are few important considerations while using the universal logic gates

1. To realize any logic, use the minimum number of NAND or NOR gates.
2. Don't have cascade logic as it increases the overall propagation delay.
3. Use the concept of the optimization to have the parallel logic.
4. The logic should have minimum power and maximum speed according to requirements.

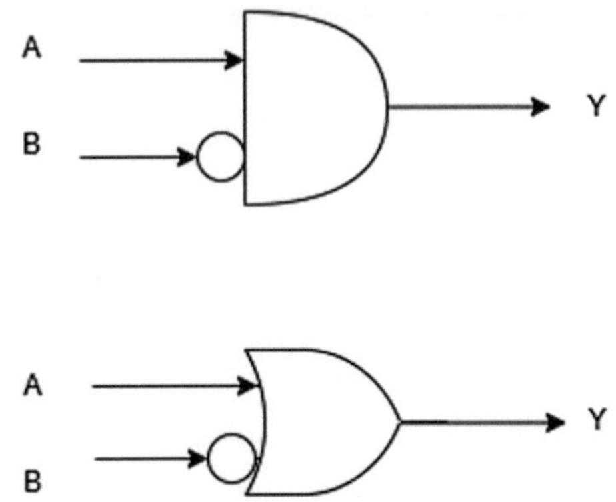

Fig. 1.18 Puzzle_Chapter_1

5. Focus on the low power aware designs and architecture while designing the digital systems.

1.8 Important Highlights

Following are few of the important points to summarize this chapter.

1. The objective of digital designer is to have the logic using minimum logic gates.
2. The area, speed, and power are important considerations in the design.
3. Most of the time digital system demands the lesser area, lesser power and maximum speed.
4. Avoid the cascading of the stages to reduce the propagation delay.
5. Use the speed optimization techniques and low power aware RTL design.
6. NAND, NOR are universal logic gates and used to implement any logic.

1.9 Puzzle

Are the following gates are used as a universal logic gates? Try this! (Fig. 1.18).

Chapter 2
Area Optimized Designs

For the complex SoC and IP design one of the main constraints is area. The number of the logic cells in the design is the area and as we know that the area constraints are limitations or restrictions placed on the physical size of the design during the ASIC design process. They define the maximum area that the design can occupy on the silicon die. These constraints are used to ensure that the final design fits within the available space and meets performance requirements. They are especially important in applications where space is limited, such as processing, computing. By considering all the above points the chapter is useful to understand the resource utilization, performance improvement by limiting the area, the clock mux logic, and the practical scenarios and use of the universal logic during the design.

2.1 Design Using NAND Gate

The NAND is universal logic gate and used to implement the any kind of Boolean function. The objective of the designer is to have the minimum number of NAND gates in the design. Now consider design of the 2-input XOR gate using the 2-input NAND gate (Figs. 2.1 and 2.2).

Step 1: Let us have the XOR using the AND-OR gates.

Step 2: We know that bubbled OR is NAND, so let us place bubble that is NOT at the inputs of OR gate. To cancel the effect of the bubble let us place bubble at the output of the AND gate.

Step 3: Sketch the design using 2-input NAND gate for the above representation.

Step 4: Design uses the five 2-input NAND gates. If we need to have the 10,000 XOR gates using NAND then the design needs area of 50,000 NAND gates.

V. Taraate, *Advanced Digital Design Techniques*,
https://doi.org/10.1007/978-981-95-4510-0_2

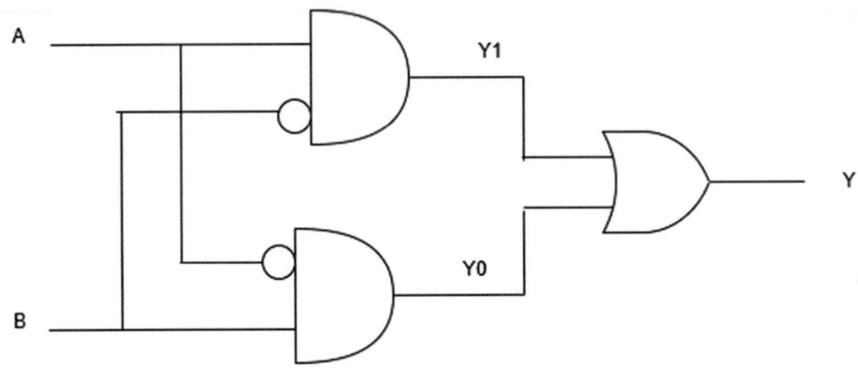

Fig. 2.1 XOR gate logic using custom gates and OR gate

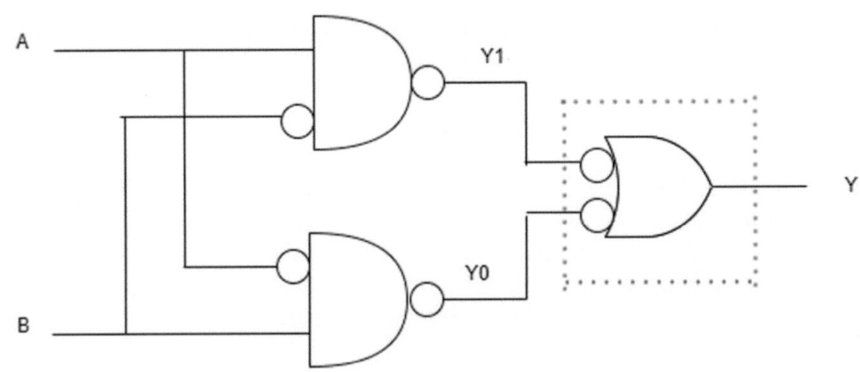

Fig. 2.2 XOR gate using the NAND and custom gates

Step 5: Optimize the design using minimum number of NAND gates. At the input side we don't need to have the two NOT gates and we can have the complement of A, complement of B using one NAND gate as complement of A OR complement of B is equivalent to NAND of A, B (Figs. 2.3 and 2.4).

Let us get the Boolean equation for Y

$$
\begin{aligned}
y2 &= \overline{AB} \\
y0 &= \overline{A.\overline{AB}} = \overline{A.(\overline{A}+\overline{B})} = \overline{A.\overline{B}} \\
y1 &= \overline{B.\overline{AB}} = \overline{B.(\overline{A}+\overline{B})} = \overline{B.\overline{A}} \\
Y &= \overline{y0.y1} = \overline{\overline{(A.\overline{B})}.\overline{(B.\overline{A})}}.
\end{aligned}
$$

Using De Morgan's theorem that is NAND is equal to bubbled OR we will get $Y = \overline{\overline{(A.\overline{B})}.\overline{(B.\overline{A})}}$

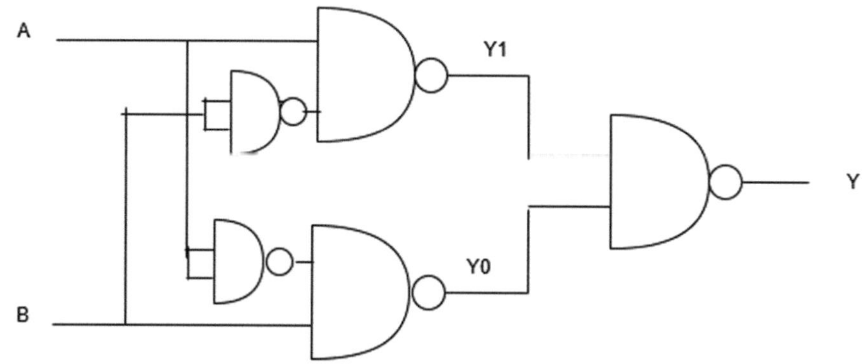

Fig. 2.3 XOR gate using the NAND gates

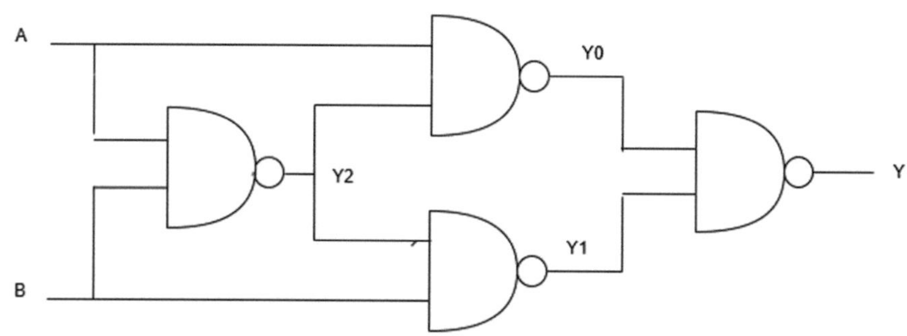

Fig. 2.4 Logic realization of XOR using minimum number of NAND

$$Y = \overline{\overline{(A.\overline{B})}} + \overline{\overline{(B.\overline{A})}}$$
$$Y = A.\overline{B} + \overline{A}.B.$$

As discussed above the two input XOR gate needs only four 2-input NAND gates and hence for the 10,000 XOR gate we need to have the 40,000 NAND gates. So as compared to the non-optimized logic the optimized logic area is lesser by 10,000 gates.

2.2 The Multiplexer as a Universal Logic

The multiplexer is universal logic and referred as mux and has many inputs and single output. To select one of the inputs it uses the select line. It is used as data selector element and we can design any combinational logic using the minimum number of multiplexers (Fig. 2.5 and Table 2.1).

As shown for the SEL = 0 the Y is I0 and for the SEL = 1 the Y = I1.

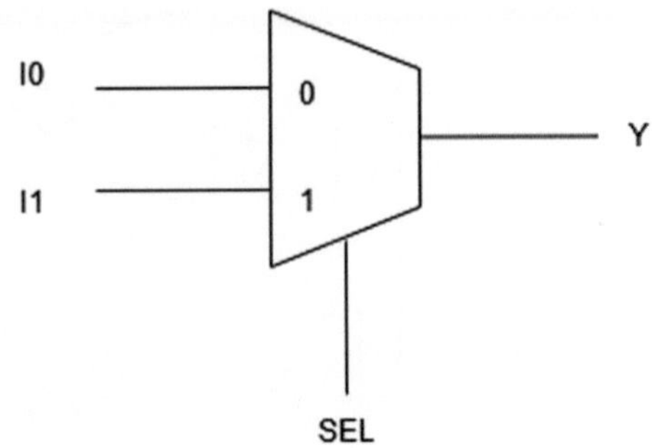

Fig. 2.5 2:1 mux

Table 2.1 Truth-table of mux

SEL	Y	Product term
0	I0	$\overline{SEL}.I0$
1	I1	$SEL.I1$

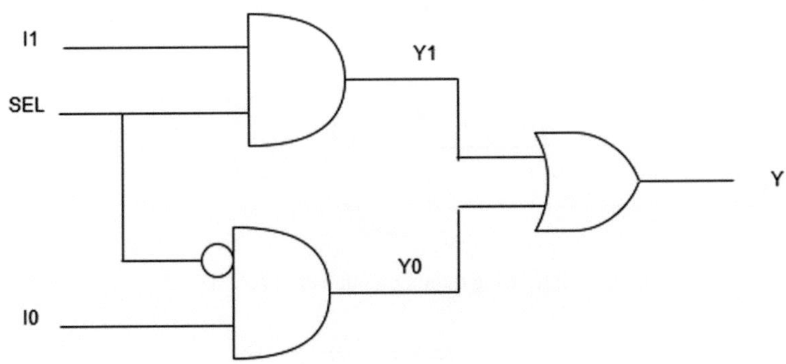

Fig. 2.6 2:1 mux

The sum of product (SOP) expression is

$$Y = SEL.I1 + \overline{SEL}.I0.$$

That means for the SEL = 1 Y = I1 and for the SEL = 0 Y = I0 (Fig. 2.6).

2.3 The 2:1 Mux Using the Minimum Number of NAND Gates

As discussed in the above section the 2:1 mux is realized using the AND-OR-NOT gates (Figs. 2.7, 2.8 and 2.9).

To design the 2:1 mux using NAND gates use the following steps.

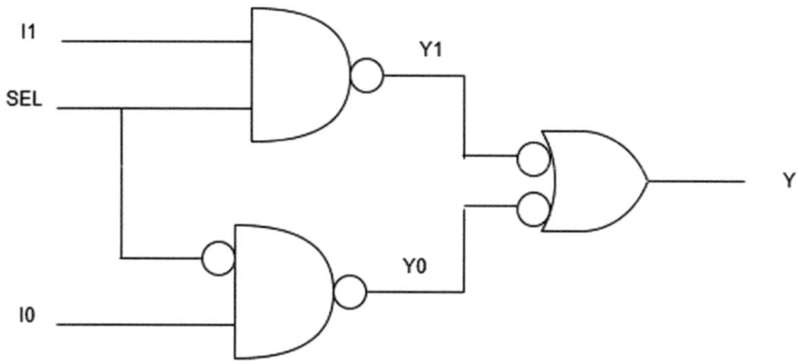

Fig. 2.7 2;1 MUX using bubbled OR, NAND

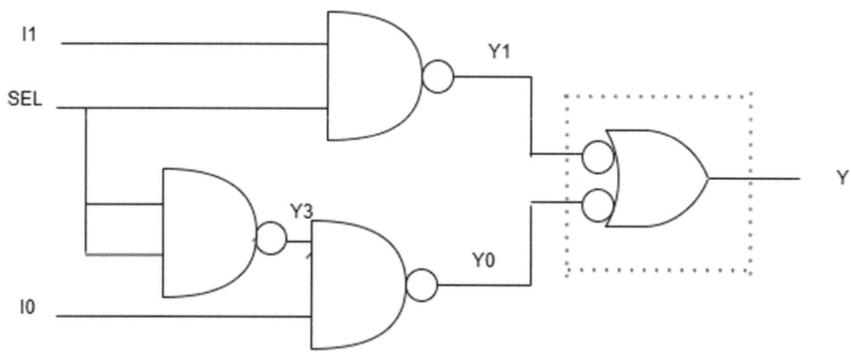

Fig. 2.8 2:1 mux using the min gates

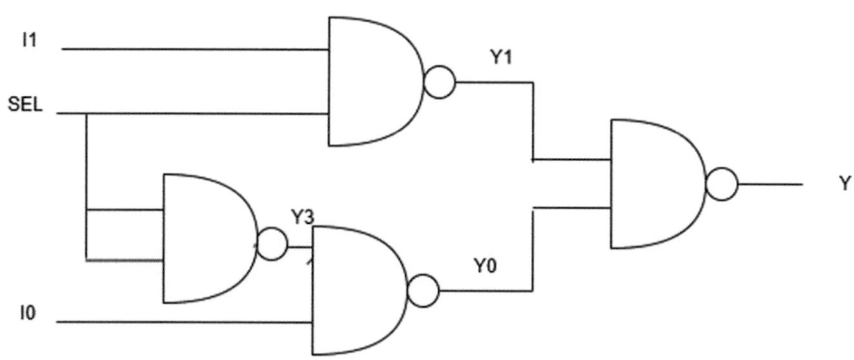

Fig. 2.9 2:1 mux using NAND gates

Table 2.2 8-bit adder inputs and outputs

Operands/outputs	Description
A	The 8-bit binary input to adder
B	The 8-bit binary input to adder
SUM	The 8-bit binary output from the adder
CARRY	The single bit carry output

Step 1: We know that bubbled OR is NAND, so place bubble(NOT) at each input of OR and to cancel the effect of the bubble place bubble at output of each AND.

Step 2: Now let us resketch the diagram to replace bubbled OR by NAND gate.

Step 3: Replace the bubbled OR using NAND. We need 4 NAND gates to implement the 2:1 mux.

2.4 Design of the n-Bit Adder

The design of the n-bit adder using the component as a full-adder is shown below. Consider $n = 8$ and we need to have the 8 full-adders. The operation performed is A + B (Table 2.2).

What we need to do is

1. Use the full-adder as a component.
2. Pass the carry output from the lower stage to the next subsequent stage.
3. Have the 8 full-adders. To optimize area LSB full adder can be replaced by half adder.
4. Get the output SUM (8-bit) and carry out (CARRY 1-bit).

The design is shown in Fig. 2.10.

2.5 Design of the n-Bit Subtractor

The design of the n-bit subtractor using the component as a full-subtractor is shown below. Consider $n = 8$ and we need to have the 8 full-subtractors. The operation performed is A–B (Table 2.3).

What we need to do is

1. Use the full-subtractor as a component.
2. Pass the borrow output from the lower stage to the next subsequent stage.
3. Have the 8 full-subtractors.
4. Get the output DIFFERENCE (8-bit) and borrow out (BORROW 1-bit)

Fig. 2.10 8-bit full-adder

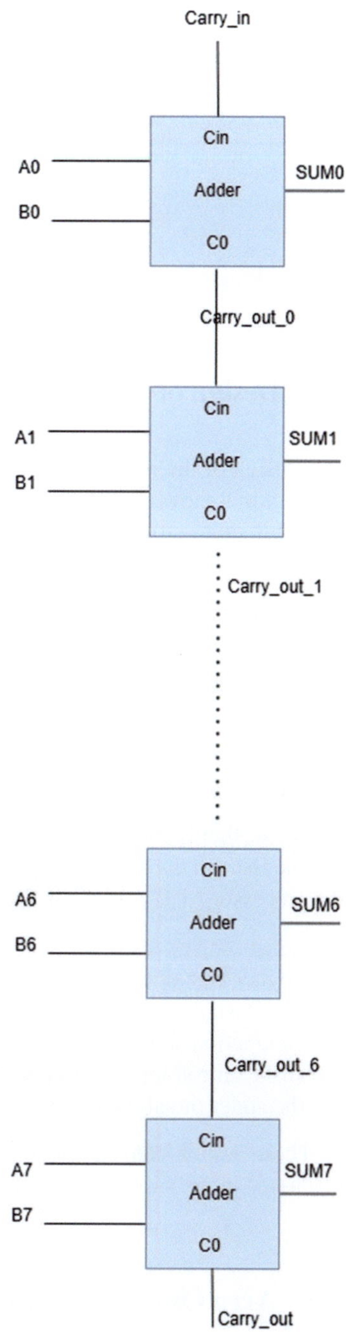

Table 2.3 8-bit subtractor inputs and outputs

Operands/outputs	Description
A	The 8-bit binary input to subtractor
B	The 8-bit binary input to subtractor
Difference	The 8-bit binary output from the subtractor
Borrow	The single bit borrow output

The design is shown in Fig. 2.11.

2.6 Design of the n-Bit Adder-Subtractor

The design of the n-bit adder-subtractor using the components as a full-adder, full-subtractor is shown below. Consider $n = 8$ and we need to have the 8 full-adders and 8-bit full-subtractors. The operation performed is A + B for Control_Input = 0 and A-B for the Control_Input = 1 (Table 2.4).

What we need to do is

1. Use the 8-bit adder to perform the addition.
2. Use the 8-bit subtractor to perform the subtraction.
3. Use the multiplexer to select the SUM, DIFF depending on the Control_input.
4. Use the multiplexer to select the BORROW, CARRY depending on the Control_ input.

The design is shown in Fig. 2.12.

The issue in the above design is use of many resources. If this logic is deployed in the processor ALU to perform the addition or subtraction then we will have following issues:

1. Higher area due to use of the adders and subtractors.
2. The design performs both the operations at a time that is 8-bit addition and 8-bit subtraction so more power.
3. Just Control_input is used at the output stage multiplexers to select the output of the adder or subtractor so not an efficient design.

These issues can be resolved by using the resource sharing techniques and discussed in the subsequent section.

2.7 Area Optimized Design

To optimize the area the following steps we can follow:

Fig. 2.11 8-bit full-subtractor

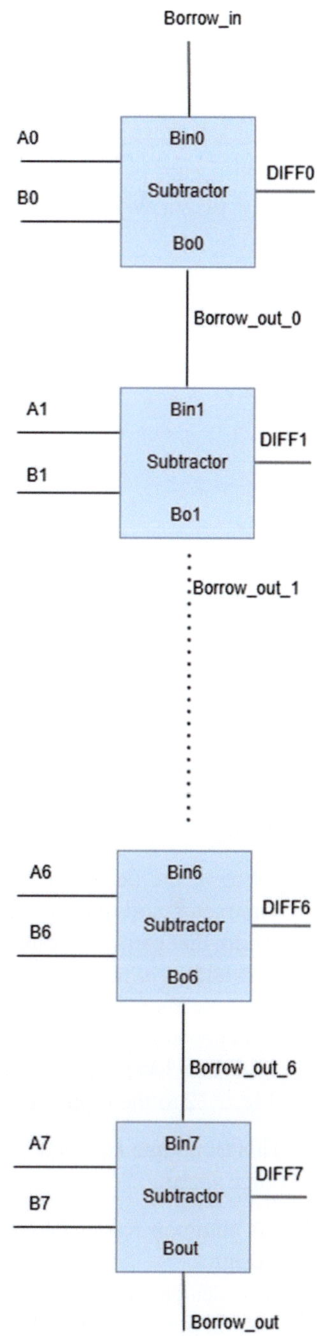

Table 2.4 8-bit adder-subtractor inputs and outputs

Operands/outputs	Description
A	The 8-bit binary input to adder/subtractor
B	The 8-bit binary input to adder/subtractor
ADD_SUB_OUT	The 8-bit binary output from the adder/subtractor
CARRY_BORROW_OUT	The single bit carry/borrow output

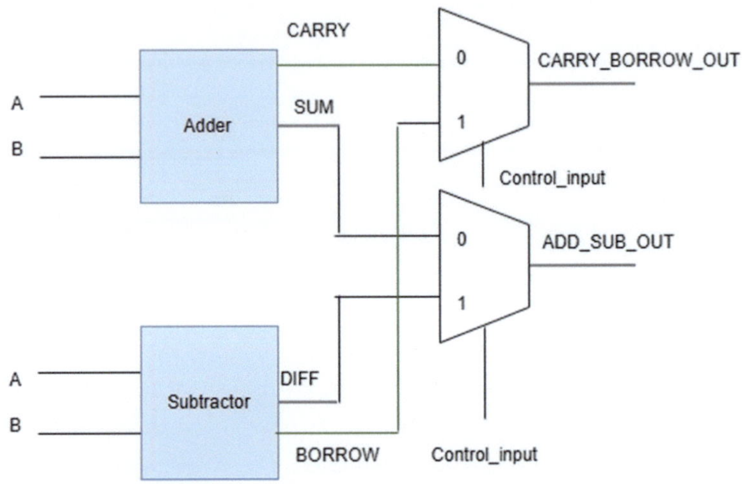

Fig. 2.12 8-bit adder-subtractor

1. Observe the design resources, in the above example the resources are adder, subtractor, and multiplexers.
2. Think that can we implement the subtraction using adder? Answer is yes.
3. Use adders to perform the subtraction. Subtraction is 2's complement addition.
4. That means for the n-bit adder-subtractor we need to deploy only n-bit adders. And depending on the Control_input pass the inputs to the adders.
5. Now how? We can push the output stage that is multiplexer chain shown in the Fig. 2.12 to the input side and adder as common resource to the output side.

This technique is called as resource sharing and following are advantages of the resource sharing:

1. As common resources are shared the design has lesser area and it is optimized design
2. The design performs only one operation at a time that is either addition or subtraction and hence the lesser power.

Figure 2.13 gives the details of the design with the resource sharing concept (Table 2.5).

Fig. 2.13 Adder-subtractor using the resource sharing

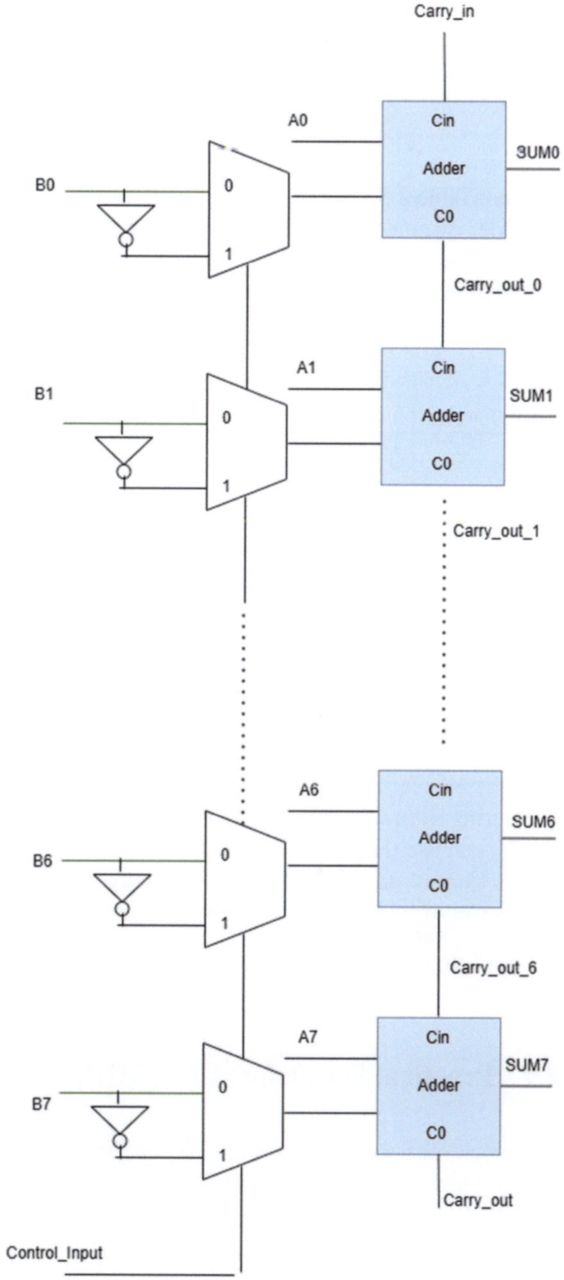

Table 2.5 Operational table for the addition/subtraction

Control_input	Operation
0	A + B
1	A–B = A + ~ B + 1

From Table 2.5 it is clear that during the addition/subtraction operation A input to the adder is same. But for the addition another input of adder is B and of subtractor it is complement of B (that is ~ B). So, we can have modified Table 2.6 with the details of the inputs and outputs as illustrated below.

Table 2.6 Modified operational table for the addition/subtraction

Control_input	Input_1 Adder	Input_2 of Adder	Carry_input	Operation
0	A	B	0	A + B + 0
1	A	~ B	1	A – B = A + ~ B + 1

2.8 Design Optimization and Resource Sharing

As discussed above, we have used the resource sharing for the area optimization and the design shown in the Fig. 2.13 can be optimized using the XOR gate.

1. During subtraction we need to have one of the inputs of the adder as ~ B.
2. We can use the Control_input XOR with the B to get ~ B.
3. As Control_input = 1 for the subtraction the XOR gate output is ~ B.
4. As Control_input = 0 for the addition the XOR gate output is B.

The design is shown in Fig. 2.14.

2.9 Practical Scenarios: CLK MUX

The clock mux is shown in Fig. 2.15 and it has two clock inputs CLK_0, CLK_1 and to select one of the input the CLK_SELECT is used. The output of the clk mux is CLK_OUT. Practically in the ASIC design we can have the clock gating cells and used for switching from faster to slower clock during the test mode.

As shown in Table 2.7, the design has CLK_OUT and for the CLK_SEL = 1 the CLK_OUT = CLK_1, for the CLK_SEL = 0 the CLK_OUT = CLK_0.

The logic of the clock mux is shown below and timing sequence in Fig. 2.16 and is prone to the glitches due to the asynchronous clock select input. And it is not recommended to use this type of clock muxing in the design.

Fig. 2.14 Adder-subtractor with optimized resources

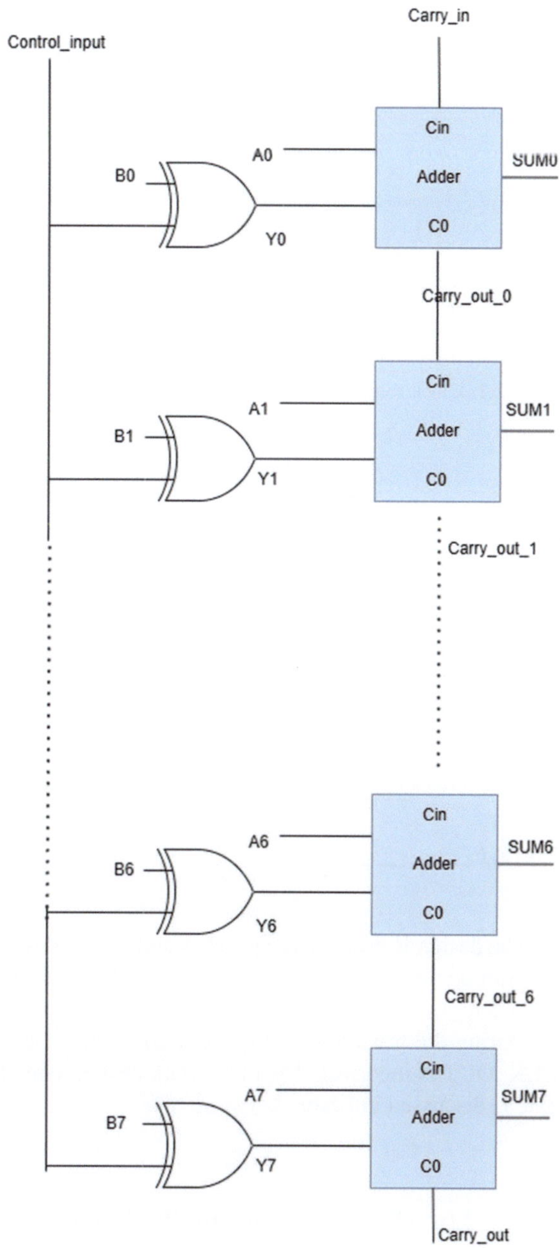

Fig. 2.15 Clock mux logic

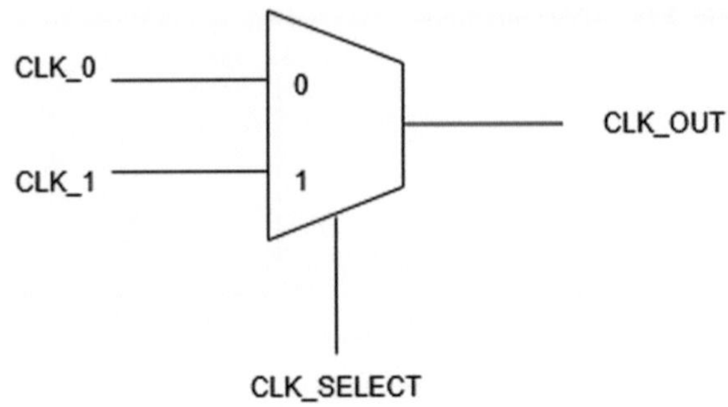

Table 2.7 Clock selection table

CLK_SEL	CLK_OUT
0	CLK_0
1	CLK_1

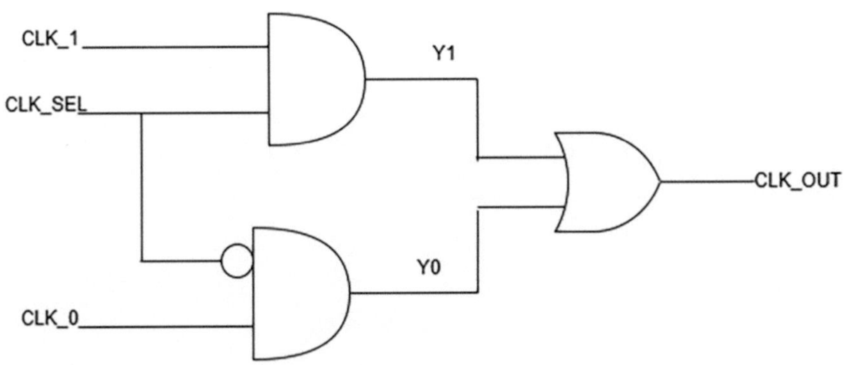

The design is prone to the glitches and we need to have the glitch free clock mux. The timing diagram is shown in Fig. 2.16. Here the CLK_SEL is asynchronous input and CLK_1, CLK_0 are two inputs of the clk mux.

During the transition of the CLK_SEL the CLK_OUT has glitches and output CLK_OUT is incorrect. The clock mux design using the synchronizers and the OR gate is discussed in Chap. 7.

2.10 Mux-Based Chain for the Selection

Now consider that in the design I need to have the data path B output (B7 to B0) or the complement of the data path output (complement of B7 to complement of B0). For the control_input = 1 the output Y should be complement of the B7 to B0 and for the Control_input = 0 the output Y is equal to the B7 to B0.

The table is shown below (Fig. 2.17).

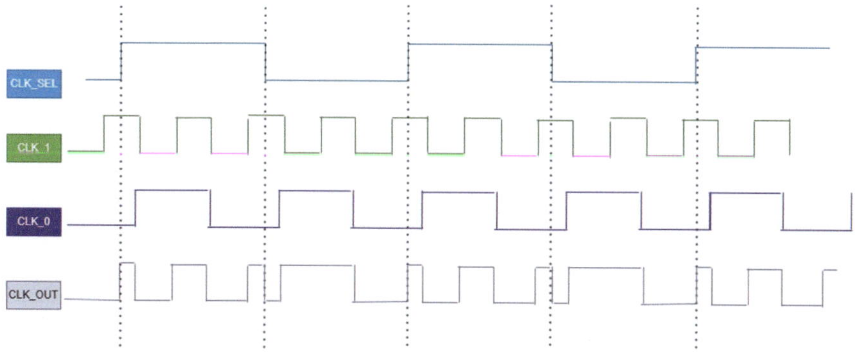

Fig. 2.16 Timing diagram of clock mux

Fig. 2.17 Functional truth-table of the design

Control_Input	Y7	Y6	Y5	Y4	Y3	Y2	Y1	Y0
0	B7	B6	B5	B4	B3	B2	B1	B0
1	$\overline{B7}$	$\overline{B6}$	$\overline{B5}$	$\overline{B4}$	$\overline{B3}$	$\overline{B2}$	$\overline{B1}$	$\overline{B0}$

The design has the 8 outputs $Y7$ to $Y0$ where $Y7$ is MSB and $Y0$ is LSB and needs the eight 2:1 multiplexer as a data path selector element and the eight NOT gates to get complement of the B7 to B0. The design is shown in Fig. 2.18.

2.11 The Scan Mux

In the DFT that is design for the testability; to improve the controllability and observability of the design we need to have the mux-based scan.

The debug or scan mux is shown in Fig. 2.19.

During the scan or the debug mode the scan_en = 1 and scan_mux_out = debug_in, in the normal mode the scan_en = 0 and scan_mux_out = data_in.

The DFT friendly designs are discussed in Chap. 11.

2.12 Important Highlights

Following are few of the important points to summarize this chapter.

Fig. 2.18 Mux-based chain at output

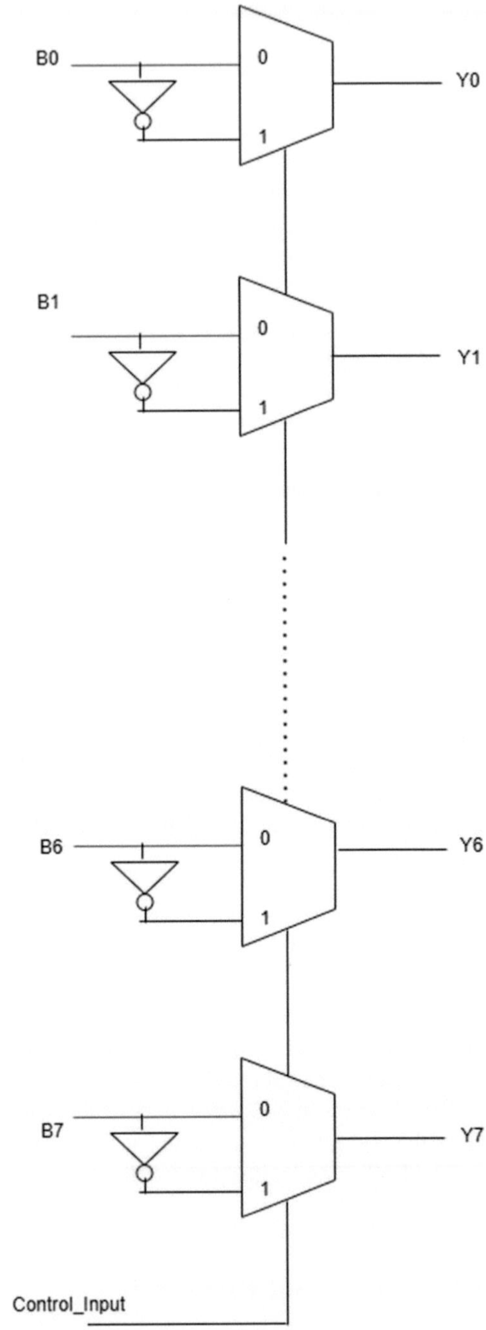

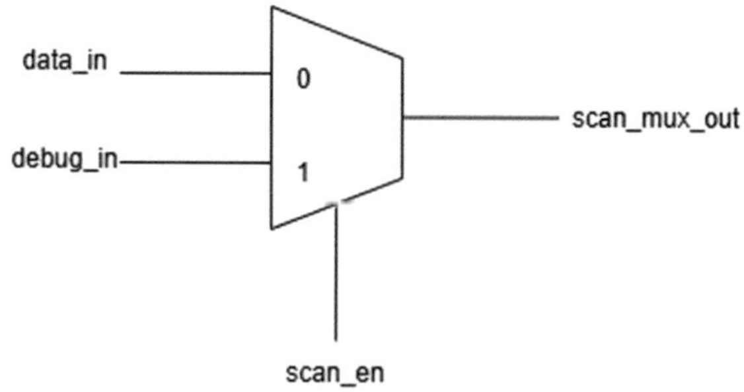

Fig. 2.19 Debug or scan mux

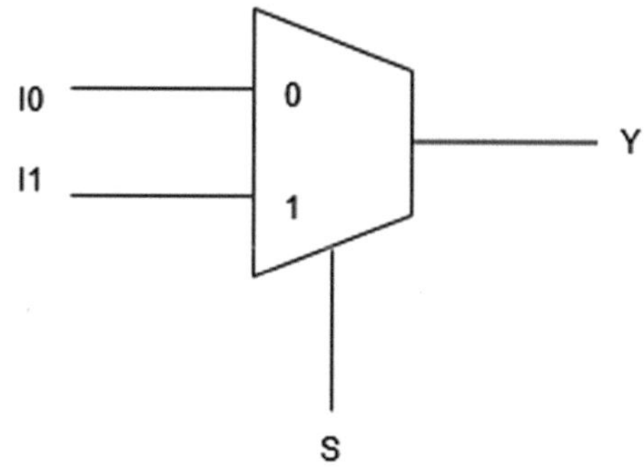

Fig. 2.20 Puzzle_Chapter_2

1. The area constraints are limitations or restrictions placed on the physical size of the design during the ASIC design process.
2. The area constraints define the maximum area that the design can occupy on the silicon die. These constraints are used to ensure that the final design fits within the available space and meets performance requirements.
3. The area constraints are especially important in applications where space is limited, such as processing, computing.
4. The resource sharing is one of the techniques to optimize the area
5. Using resource sharing as common resources are shared the design uses the lesser area and lesser power.
6. The mux is universal logic and used to design any combinational logic
7. The mux is used to have the clock muxing and also used as data selectors.

2.13 Puzzle

The following logic is universal logic ? Can you implement all the logic gates using the minimum number of multiplexers? Try this! (Fig. 2.20).

Chapter 3
The Data and Control Paths

The design has the data paths and control paths. Always it is recommended to have the separate data path and control path in the design to have the clean timing. The area optimized data path and area optimized state machines to generate the control and timing signal is the main objective of the designer. In this context the chapter is useful to understand the data and control paths in the design and how we can use the area optimization techniques while passing data from the data path for the various control signals. The chapter also covers the details of the optimization for the data and control path and use of the universal logic elements which is multiplexers and design using multiplexers.

3.1 Data Path

The data path is the hardware which is used to perform the operations on the data. Few of the operations are arithmetic, logical, and the data transfer operations.

For example, consider the pipelined processor design we can have the following element in the data path.

1. Arithmetic and logic unit (ALU).
2. Registers.
3. Buses and bus interfaces.

As discussed in Chap. 2, if we need to perform the addition and subtraction operation using the resource sharing techniques then we have the following design (Fig. 3.1).

The data path is from the input to the output through the adder as a resource. Here the Control_input is used to generate the required output to perform either the addition or subtraction. For the Control_input = 0 the design performs addition and for the Control_input = 1 the design performs the subtraction.

V. Taraate, *Advanced Digital Design Techniques*,
https://doi.org/10.1007/978-981-95-4510-0_3

Fig. 3.1 Data and control paths of adder

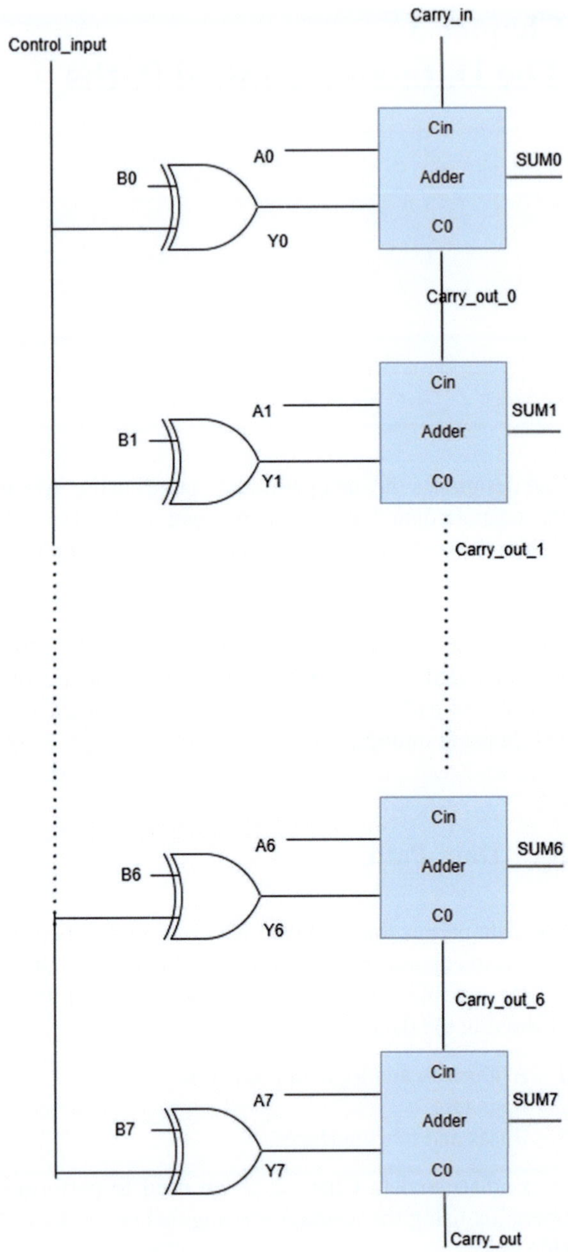

So, effectively the data path elements are used to perform specific operations on the data and to move the data from one of the functional units to another unit.

3.2 Control Path

The control path has control elements and the sequential logic which is used to generate the control or timing or both the control and timing signals to the design which is in the data path.

The control and timing unit are used to provide the control and timing inputs and it determines which operation the data path should perform and how the data should be routed.

For any kind of the design the control unit is designed by using the FSM with the objective to generate the clean timing and control signals.

The sequential designs and FSM control designs are subsequently discussed in the next few chapters. In this chapter let us discuss the basic control elements and their use and the area optimization schemes which we need to deploy.

As shown in Fig. 3.2 we have the control element as a multiplexer. And the Control_in generated from timing and control is used to generate the B or ~ B for the 0 or 1 respectively. Hence, we can call this as a control element as it controls the data to be processed from B0 to B7 to one of the inputs of the adder.

Let us discuss the multiplexer as a data control element and also it is called as universal logic to implement the combinational design.

We have two types of the design: combinational design has output which is function of the present input only. Sequential design has an output which is function of the present input and past output.

3.3 Multiplexer as a Control Element

Consider the multiplexer 2:1, and also referred as mux. Mux has two inputs and to select one of the inputs at a time it needs select line (select input) as one. It is also referred as the many to one switch and used in the various designs. For example, clock muxing, data selector, switch, to perform the logical operations.

The following section describes the use of the mux as a universal logic to implement various Boolean functions and logic gates.

Fig. 3.2 Adder-subtractor design

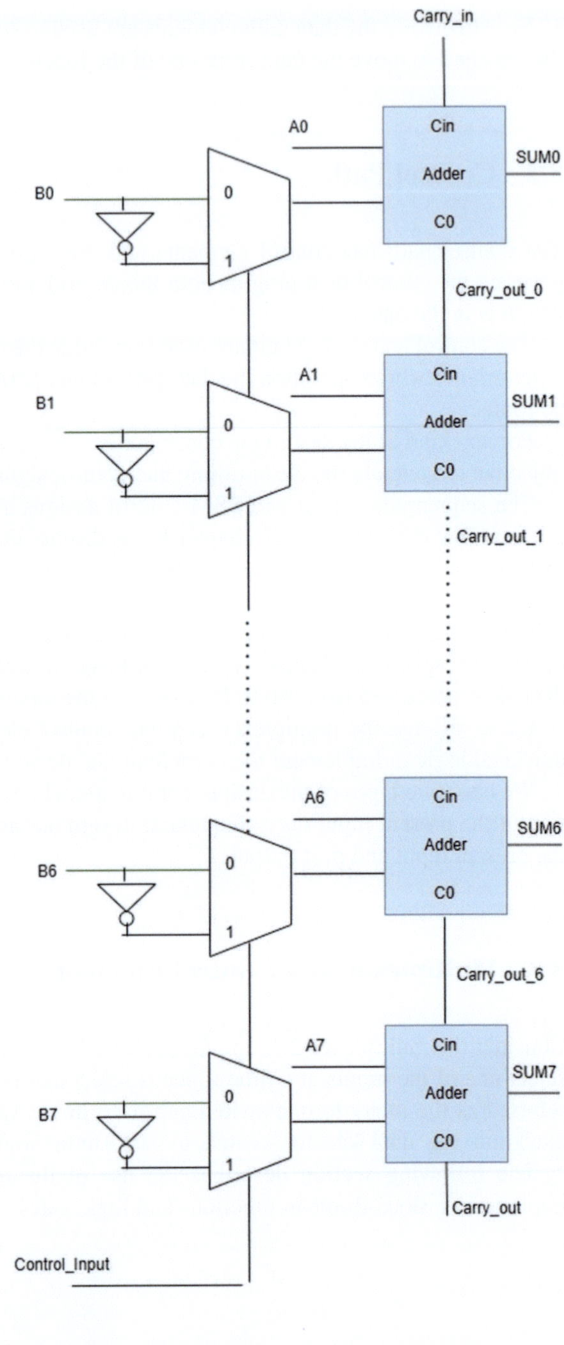

Table 3.1 Complement logic

A	$Y = \overline{A}$
0	1
1	0

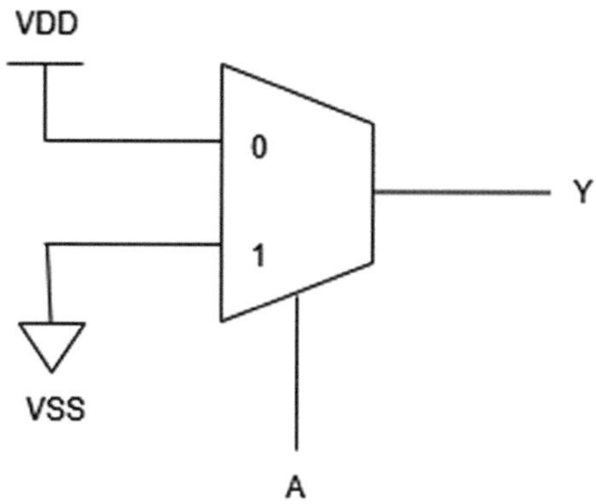

Fig. 3.3 Complement logic using 2:1 mux

3.3.1 *Complement Logic*

As we know that complement of 1 is 0 and complement of 0 is 1. That is there is 180-degree phase shift between the input and output. Consider the NOT logic which is complement logic and described in Table 3.1.

The above logic is implemented by using the 2:1 multiplexer as shown in Fig. 3.3.

3.3.2 *AND Logic*

The 2-input AND logic is described in Table 3.2.

The optimized Table 3.3 and it is derived from Table 2.2. For the A = 0 the output Y = 0 and for A = 1 the output Y = B.

The design is implemented using the single 2:1 mux and shown in Fig. 3.4. As shown, the input A is used as select input and when it is 1 it passes logic 0 to the output Y. When A = 1 then it passes B to the output.

Table 3.2 Truth-table of 2-input AND

A	B	$Y = A.B$
0	0	0
0	1	0
1	0	0
1	1	1

Table 3.3 Optimized table of 2-input AND logic

A	$Y = A.B$
0	0
1	B

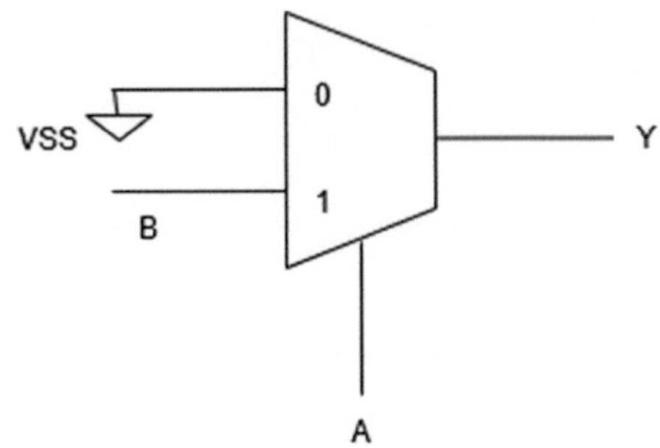

Fig. 3.4 AND logic using 2:1 mux

3.3.3 *NAND Logic*

The 2-input NAND logic is described in Table 3.4.

The optimized Table 3.5 and it is derived from Table 3.4. For the A = 0 the output Y = 1 and for A = 1 the output $Y = \overline{\boldsymbol{B}}$.

The design is implemented using the single 2:1 mux and NOT logic and shown in Fig. 3.5. As shown, the input A is used as select input and when it is 1 it passes logic 1 to the output Y. When A = 1 then it passes complement of B to the output. Complement is denoted by ~.

The design of NAND logic using only 2:1 mux is shown in Fig. 3.6 NAND logic using 2:1 and uses two 2:1 multiplexer. Here NOT logic is replaced by the 2:1 multiplexer.

Table 3.4 Truth-table of 2-input NAND

A	B	$Y = \overline{AB}$
0	0	1
0	1	1
1	0	1
1	1	0

Table 3.5 Optimized table of 2-input NAND logic

A	$Y = \overline{AB}$
0	1
1	$\overline{B}$

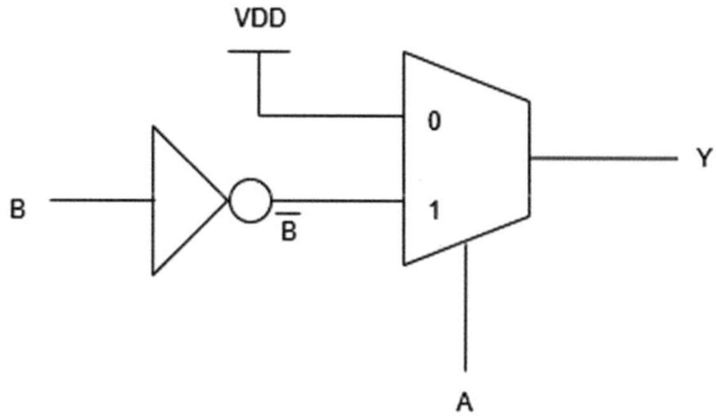

Fig. 3.5 NAND logic using 2:1 mux

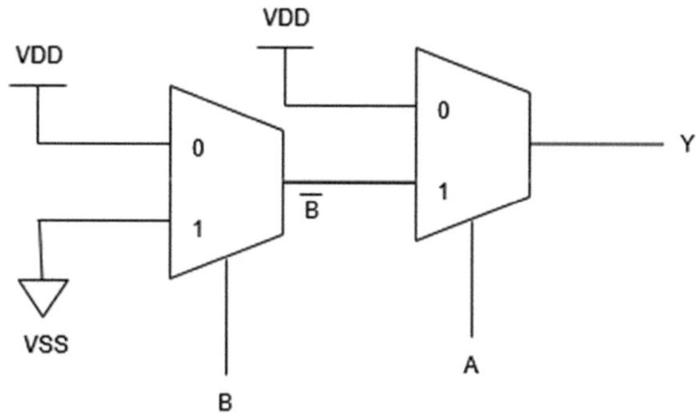

Fig. 3.6 NAND logic using 2:1 multiplexer

3.3.4 OR Logic

The 2-input OR logic is described in Table 3.6.

The optimized Table 3.7 and it is derived from Table 3.6. For the A = 0 the output Y = B and for A = 1 the output Y = 1.

Table 3.6 Truth-table of 2-input OR

A	B	$Y = A + B$
0	0	0
0	1	1
1	0	1
1	1	1

Table 3.7 Optimized table of 2-input OR logic

A	$Y = A + B$
0	B
1	1

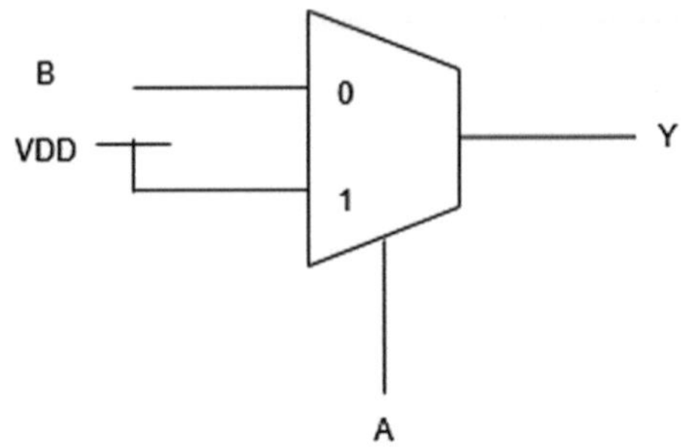

Fig. 3.7 OR logic using 2:1 mux

The design is implemented using the single 2:1 mux and shown in Fig. 3.7. As shown, the input A is used as select input and when it is 1 it passes 1 to the output Y. When A = 0 then it passes B to the output.

3.3.5 NOR Logic

The 2-input NOR logic is described in Table 3.8.

The optimized Table 3.9 and it is derived from Table 3.8. For the A = 0 the output $Y = \overline{B}$ and for A = 1 the output $Y = 0$.

The design is implemented using the single 2:1 mux and NOT logic and shown in Fig. 3.8. As shown, the input A is used as select input and when it is 0 it passes complement of B to the output Y. When A = 1 then it passes logic 0 to the output.

The design of NOR logic using only 2:1 mux is shown in Fig. 3.9 NOR logic using 2:1 and uses two 2:1 multiplexer. Here NOT logic is replaced by the 2:1 multiplexer.

Table 3.8 Truth-table of 2-input NOR

A	B	$Y = \overline{A+B}$
0	0	1
0	1	0
1	0	0
1	1	0

Table 3.9 Optimized table of 2-input NOR logic

A	$Y = \overline{A+B}$
0	$\overline{B}$
1	0

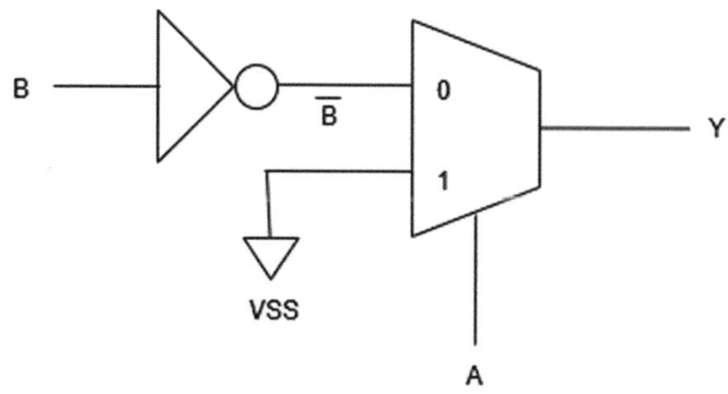

Fig. 3.8 NOR logic using 2:1 mux

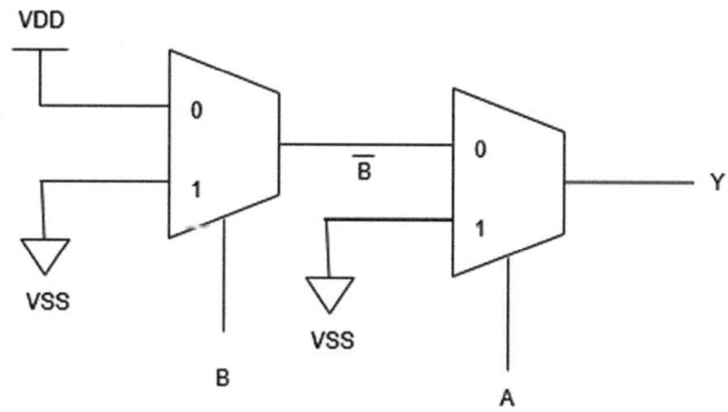

Fig. 3.9 NOR logic using 2:1 multiplexer

3.3.6 *XOR Logic*

The 2-input XOR logic is described in Table 3.10.

The optimized Table 3.11 and it is derived from Table 3.10. For the A = 0 the output Y = B and for the A = 1 the output Y = $\overline{\boldsymbol{B}}$.

The design is implemented using the single 2:1 mux and NOT logic and shown in Fig. 3.10. As shown, the input A is used as select input and when it is 0 it passes B to the output Y. When A = 1 then it passes complement of B to the output.

Table 3.10 Truth-table of 2-input XOR

A	B	$Y = A.\overline{B} + \overline{A}.B$
0	0	0
0	1	1
1	0	1
1	1	0

Table 3.11 Optimized table of 2-input XOR logic

A	$Y = A.\overline{B} + \overline{A}.B$
0	B
1	$\overline{B}$

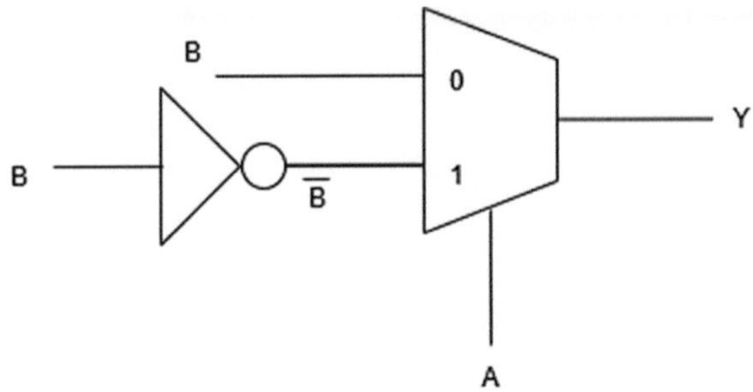

Fig. 3.10 XOR logic using 2:1 mux

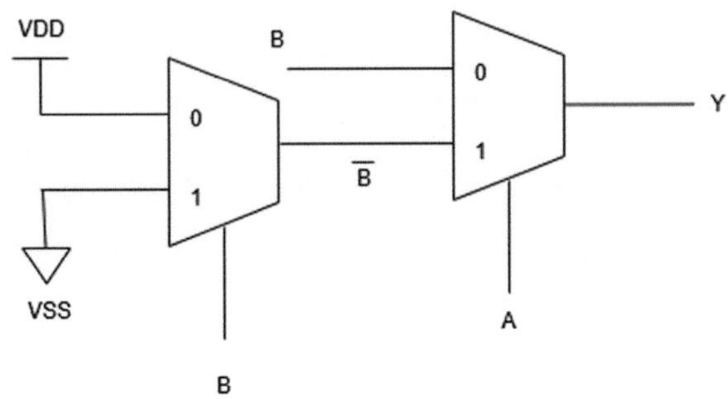

Fig. 3.11 XOR logic using 2:1 multiplexer

The design of XOR logic using only 2:1 mux is shown in Fig. 3.11: XOR logic using 2:1 and uses two 2:1 multiplexer. Here NOT logic is replaced by the 2:1 multiplexer.

3.3.7 XNOR Logic

The 2-input XNOR logic is described in Table 3.12.

The optimized Table 3.13 is shown below and it is derived from Table 3.12. For the A = 0 the output Y = $\overline{\boldsymbol{B}}$ and Y = B for the A = 1.

The design is implemented using the single 2:1 mux and NOT logic and shown in Fig. 3.12. As shown, the input A is used as select input and when it is 0 it passes complement of B to the output Y. When A = 1 then it passes B to the output.

Table 3.12 Truth-table of 2-input XNOR

A	B	$Y = \overline{AB} + A.B$
0	0	1
0	1	0
1	0	0
1	1	1

Table 3.13 Optimized table of 2-input XNOR logic

A	$Y = \overline{A}\overline{B} + A.B$
1	B
0	$\overline{B}$

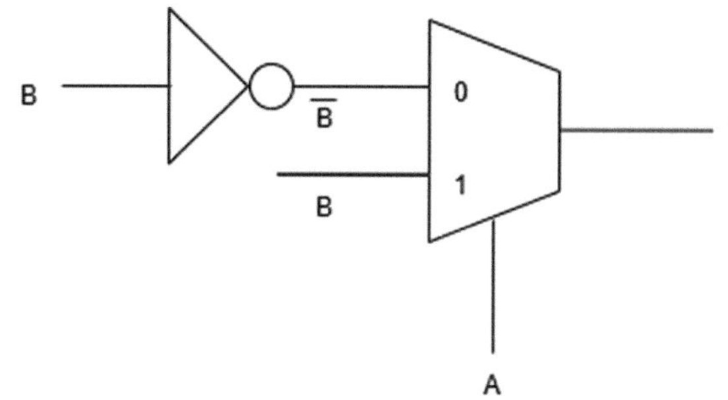

Fig. 3.12 XNOR logic using 2:1 mux

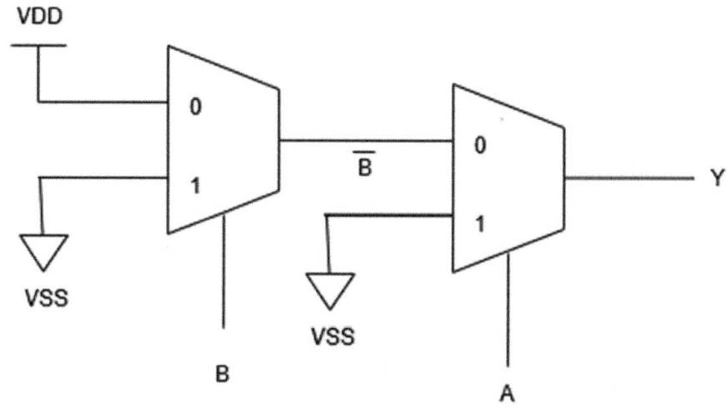

Fig. 3.13 XNOR logic using 2:1 multiplexer

The design of XNOR logic using only 2:1 mux is shown in Fig. 3.13: XNOR logic using 2:1 and uses two 2:1 multiplexer. Here NOT logic is replaced by the 2:1 multiplexer.

3.4 Universal Logic Gate

NAND and NOR are universal logic gates and we have the custom ASIC gates which are discussed in the following section which are universal gates. By using these gates, we can design any combinational logic.

Table 3.14 Truth-table of 2-input custom universal gate

A	B	$Y = A\overline{B}$
0	0	0
0	1	0
1	0	1
1	1	0

Table 3.15 Optimized table of 2-input custom universal gate

A	$Y = A\overline{B}$
0	0
1	$\overline{B}$

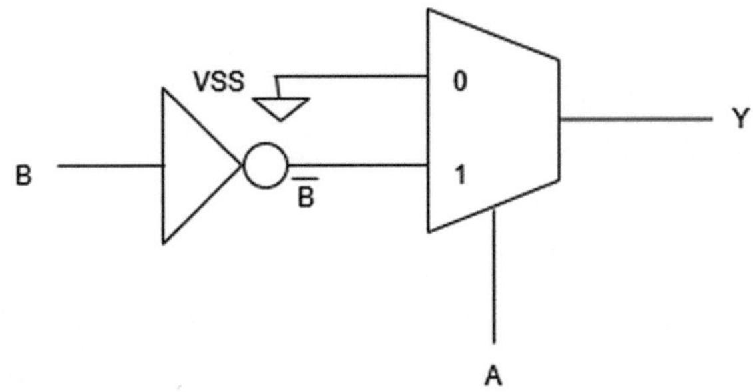

Fig. 3.14 Custom 2-input logic using 2:1 mux and NOT

3.4.1 *The Custom Universal Gate and Use in Optimization*

The 2-input custom ASIC gate is described in Table 3.14.

The optimized Table 3.15 is shown below and it is derived from Table 3.14. For the A = 1 the output Y = $\overline{\boldsymbol{B}}$ **and** Y = 0 for the A = 0.

The design is implemented using the 2:1 mux and NOT logic and shown in Fig. 3.14. As shown, the input A is used as select input and when it is 1 it passes complement of B to the output Y. When A = 0 then it passes 0 to the output.

The design of custom universal gate using only 2:1 mux is shown in Fig. 3.15: Custom logic using 2:1 and uses two 2:1 multiplexer. Here NOT logic is replaced by the 2:1 multiplexer (Fig. 3.15).

3.4.2 *The Custom Universal Gate-II and Use in Optimization*

The 2-input custom ASIC gate is described in Table 3.16.

The optimized Table 3.17 and it is derived from Table 3.16. For the A = 0 the output Y = $\overline{\boldsymbol{B}}$ **and** Y = 1 for the A = 1.

Fig. 3.15 Custom logic using 2:1 multiplexer

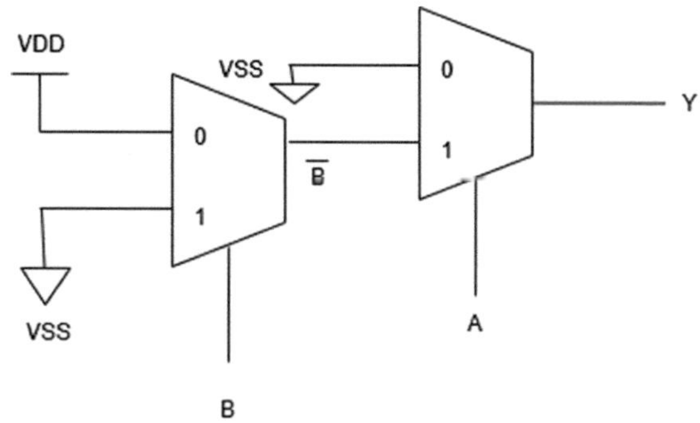

Table 3.16 Truth-table of 2-input custom universal gate-II

A	B	$Y = A + \overline{B}$
0	0	1
0	1	0
1	0	1
1	1	1

Table 3.17 Optimized table of 2-input custom universal gate-II

A	$Y = A + \overline{B}$
0	$\overline{B}$
1	1

The design is implemented using the 2:1 mux and NOT logic and shown in Fig. 3.16. As shown, the input A is used as select input and when it is 0 it passes complement of B to the output Y. When A = 1 then it passes 1 to the output.

The design of custom universal gate using only 2:1 mux is shown in Fig. 3.17: Custom logic gate-II using 2:1 and uses two 2:1 multiplexer. Here NOT logic is replaced by the 2:1 multiplexer.

Fig. 3.16 Custom gate-II 2-input logic using 2:1 mux and NOT

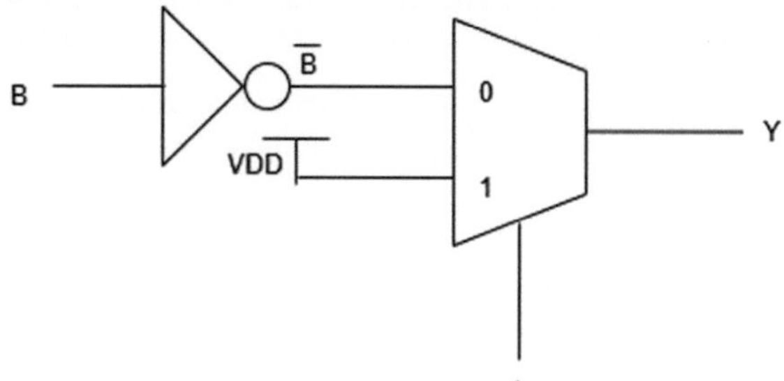

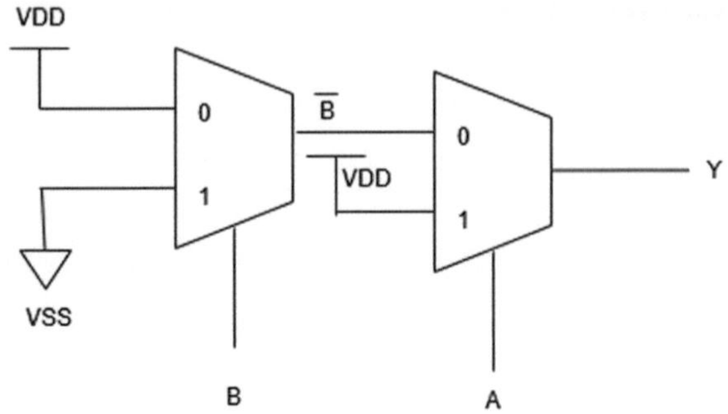

Fig. 3.17 Custom logic gate-II using 2:1 multiplexer

3.5 The Optimization Using Mux

The *n*:1 mux has *n* input lines and single output line. It has the m control inputs and also called as select inputs. The relationship between the n and m is n is equal to 2 to the power of m. It is also called as many to one switch. The multiplexer or mux is universal logic and used to design the combinational or combinatorial logic. Combinational logic is also called as combo logic.

Now consider the 4:1 multiplexer and it has $n = 4$ inputs, $m = 2$ select inputs and single output. Similarly, if we have 8:1 mux then it has the $n = 8$ inputs, $m = 3$ select inputs and single output. Depending on the select line one of the inputs is selected and passed to the output.

Consider we need to have 12:1 mux if I have 12 data paths in the design then to select one of the inputs at a time, we need to have a 4 select or control lines. These control lines are generated by the control path logic and accordingly the data passes from one of the data paths to the output of mux.

Multiplexers are widely used in the clock muxing and as a control element or data selector in digital design. The applications of the multiplexers are discussed in detail in next few chapters.

3.5.1 *The 4:1 Mux as a Data Path Selector*

The 4 to 1 mux has $n = 4$ inputs, $m = 2$ select lines, and single output line Y. Consider the design has four data paths and named as I0, I1,I2, I3 then to select the data from one of the data paths we need to have two select inputs S1, S0. The truth-table of the 4:1 mux is shown in Table 3.18.

The 4:1 mux is shown in the following Fig. 3.18: 4:1 mux and the applications of the multiplexers to have efficient combo design are discussed in Chap. 4.

Table 3.18 Truth-table of 4:1 mux

S1	S0	Y
0	0	I0
0	1	I1
1	0	I2
1	1	I3

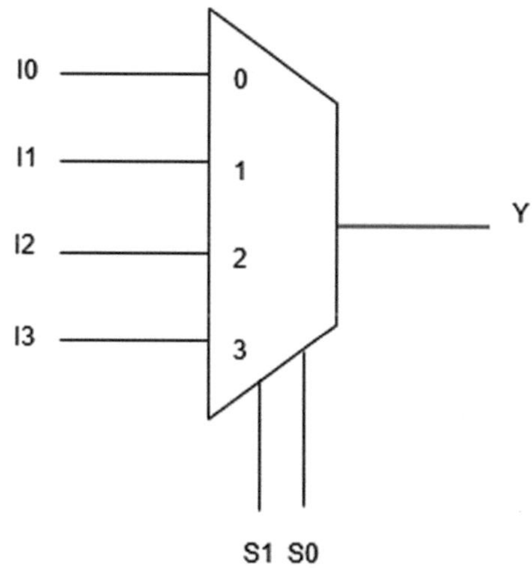

Fig. 3.18 4:1 mux

3.5.2 The Optimized Design of 4:1 Mux

The design of the 4:1 mux using three 2:1 mux is discussed in this section. Consider the following Table 3.19.

From Table 3.19 it is clear that we need to have three 2:1 mux. As we have four inputs the two 2:1 multiplexer having the select input S0 to control input mux two stages. Depending on the S0 we will get the output Y0, Y1.

1. When S0 = 0 then Y0 = I0.
2. When S0 = 1 then Y0 = I1.
3. When S0 = 0 then Y1 = I2.
4. When S0 = 0 then Y1 = I3.

Table 3.19 Truth-table of 4:1 mux using 2:1 mux

S1	S0	Y
0	0	I0
0	1	I1
1	0	I2
1	1	I3

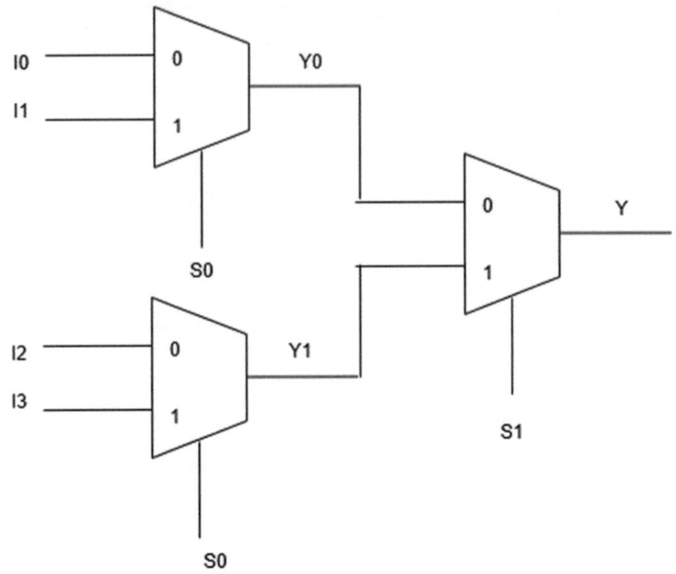

Fig. 3.19 4:1 mux using three 2:1 mux

The output stage multiplexer uses inputs as Y0, Y1 and when S1=0 the output of mux is Y0, for the S1=1 the output of mux is Y1.

The design is implemented using the three 2:1 mux and shown in Fig. 3.19.

If each mux has the propagation delay of 1ns then the overall propagation delay is 1ns + 1ns = 2ns. Here the 4:1 mux is two stage logic. During the design if we have the dedicated mux cell then depending on the technology node requirements, we can design the clock mux logic.

Consider the XOR logic, as discussed previously for A == B the output of XOR gate is logic 0. If A is not equal to B, then the XOR gate output is 1. The design of the XOR logic gate using 4:1 mux is shown in Fig. 3.20.

Here as A, B are used as select inputs the output is either 0 or 1 depending on the status of select inputs. Here A is MSB and B is LSB.

A	B	Y
0	0	VSS
0	1	VDD
1	0	VDD
1	1	VSS

The NOT of XOR is XNOR and is used as equality detector in most of the designs. Even XOR and XNOR gates are used to detect the parity and as control gates also (Fig. 3.21).

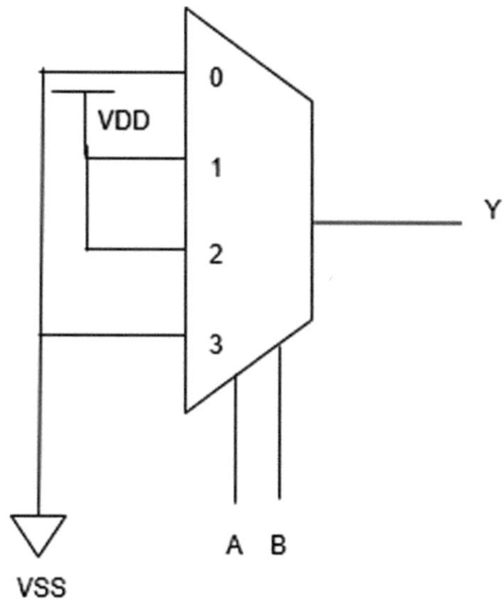

Fig. 3.20 2-input XOR using 4:1 mux

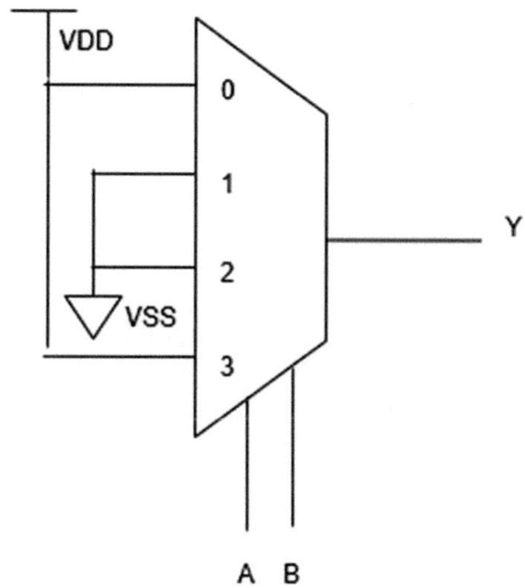

Fig. 3.21 XNOR using 4:1 mux

Here as A, B are used as select inputs the output is either 0 or 1 depending on the status of select inputs. Here A is MSB and B is LSB.

A	B	Y
0	0	VDD
0	1	VSS
1	0	VSS
1	1	VDD

3.6 Important Highlights

Following are few of the important points to summarize this chapter.

1. The multiplexer is the data selector and also referred as universal logic.
2. Few of the processor data path elements are arithmetic logic unit, registers, and the buses.
3. The multiplexers are used to implement any Boolean function or logic gates
4. The design should have separate data and control path
5. Cascade logic will incur more delay and it is not recommended to use the cascade stages in the design
6. The resource sharing should be used to optimize the design
7. The control path should provide the clean timing and control signals to the design.

3.7 Puzzle

Design the logic unit to perform the operations on the two 8-bit binary numbers A, B. Operations are NOT A, A OR B, A AND B, A XOR B. How many multiplexers and how many logic gates you need? Can you optimize the logic unit using minimum logic elements.

Chapter 4
The Area Optimization: Case Study

In the previous few chapters, we have discussed about the details of the basic combinational elements, data and control paths, even we have discussed few scenarios of the area optimization. This chapter is useful to understand the area optimization and logic design using minimum number of logic elements. For any kind of the ASIC design the least area indicates the lesser number of logic gates to implement the design functionality. By considering this as an objective the design optimization scenarios are included in this chapter.

4.1 Resource Sharing

As discussed previously the resource sharing is the powerful technique and it is used to optimize the area. The lesser logic density is the objective and consider the scenario, that I need to perform two operations on the 8-bit binary numbers A, B. The operations are addition, subtraction respectively.

As a logic designer what I will use the following strategy.

1. Use the adder to perform the addition.
2. Use the subtraction to perform the subtraction.
3. At the output stage let us have the multiplexer to select the Add, SUB output.
4. At the output stage let us have the multiplexer to select either carry or borrow.

The following is the truth-table (Table 4.1) which describes the two operations

For the Control_input = 0 the ADD_SUB_OUT is equal to the SUM, CARRY_BORROW_OUT = CARRY and for the Control_input = 1 the ADD_SUB_OUT is equal to the DIFF, CARRY_BORROW_OUT = BORROW. The design is shown in the Fig. 4.1 and uses adder, subtractor, and multiplexers.

Following are issues

1. The design performs both the operations on 8-bit binary inputs.

V. Taraate, *Advanced Digital Design Techniques*,
https://doi.org/10.1007/978-981-95-4510-0_4

Table 4.1 Truth-table of operations

Control_input	Operation
0	A + B
1	A − B

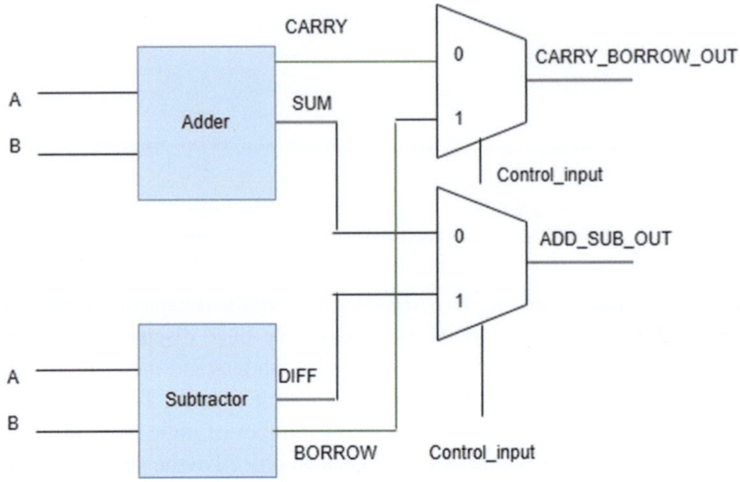

Fig. 4.1 Adder-subtractor without sharing of common resources

2. Hence more area and more power.
3. The design uses the output multiplexer chain to select either output of the adder or subtractor depending on the Control_input.
4. The design has more area without resource sharing.

As discussed in the Chap. 2, if we need to perform the addition and subtraction operation using the resource sharing techniques then we have the following design.

1. Let us use single 8-bit adder for the 8-bit operation or n-bit adder to perform the n-bit addition/subtraction.
2. As common resource adder is used the subtraction is 2's complement addition.
3. Let us modify the truth-table and let us design the logic using minimum number of logic elements.

From Table 4.2 it is clear that adder X input is always A for addition, subtraction. The adder Y input is B for the addition and complement of B for the subtraction. By considering this what we should do?

1. Use single n-bit adder as shown and let us pass A input as it is to the X input of adder.
2. Control_input should be connected to the Carry input of n-bit adder.
3. The Y input of adder is the output of data selector that is multiplexer. For the 8-bit operation we need to have eight 2:1 multiplexer.

Table 4.2 Truth-table modified to have resource sharing

Control_input	Operation X + Y + Control_input	X	Y
0	$A + B = A + B + 0$	A	B
1	$A - B = A + \overline{B} + 1$	A	$\overline{B}$

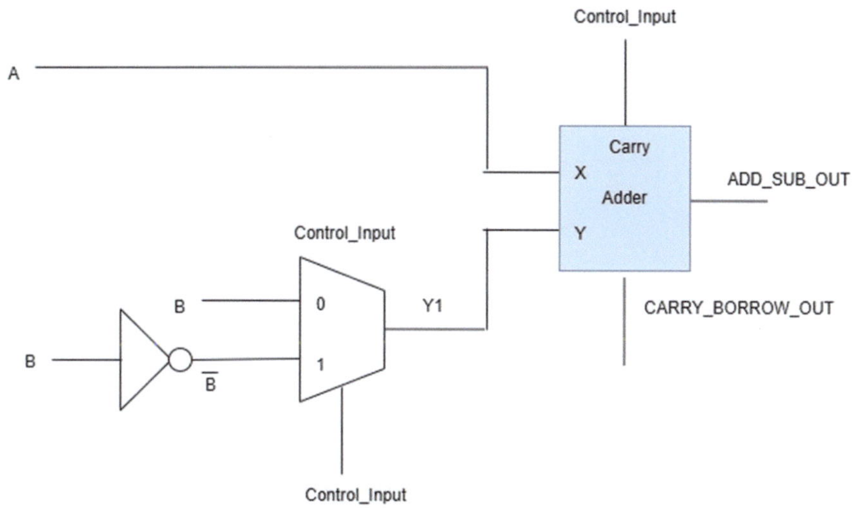

Fig. 4.2 Design using resource sharing

4. For the Control_input = 0 the output of multiplexer is B and for the Control_input = 1 the output of multiplexer is complement of B.

The design is shown in Fig. 4.2. The design uses the n-bit adder and $n \times 2{:}1$ multiplexer and n NOT gates. The design is area and power efficient and performs single operation at a time. From this we can conclude that the resource sharing technique will reduce the design resources.

The data path to have the Y input of the multiplexer can be designed using the multiplexers only. That is NOT logic is replaced by the mux. The design is shown in Fig. 4.3.

The digital logic using resource optimization to perform the addition, subtraction operation depending on the opcode is shown in Fig. 4.4. As shown, the Control_input = 0 it performs the addition and for the Control_input = 1 it performs the subtraction. The Control_input is controlled by the opcode of the instruction. For only two instructions we have single bit opcode.

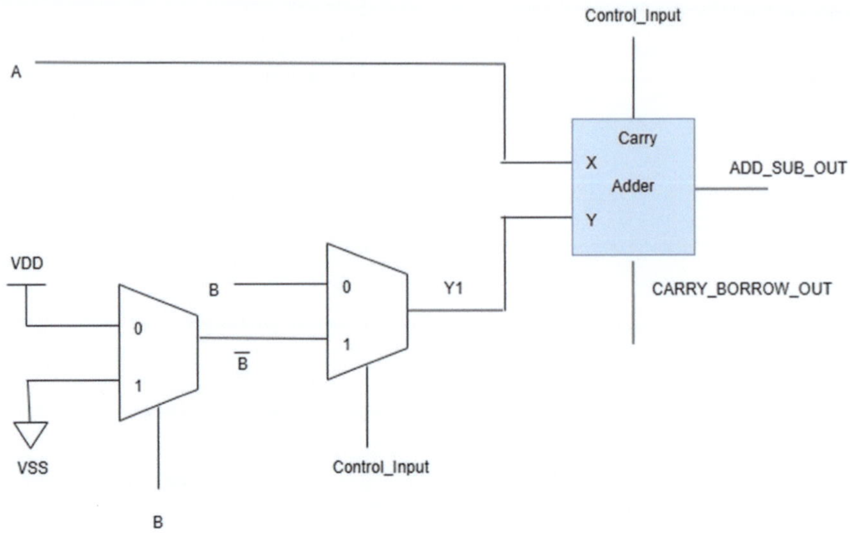

Fig. 4.3 Optimized design of the addition-subtraction

4.2 Arithmetic Unit and Area Optimization Scenarios

Now consider the design requirement is to perform the following operations on two 8-bit numbers

1. Parallel input parallel output of A.
2. Addition of A, B.
3. Subtraction of A, B, 1.
4. Decrement A.

What we need to do is following:

1. Let us have the truth-table which is also called as operational table.
2. Let us have common resource as an adder. To perform the 8-bit operations we need to have the 8 full-adders.
3. Let us include the details in the table that is adder inputs.

Opcode S1 S0	Operation	X input of adder	Y input of adder
00	$A = A + 0 + 0$	A	0
01	$A + B = A + B + 0$	A	B
10	$A - B - 1 = A + \overline{B} + 1 - 1 = A + \overline{B} + 0$	A	$\overline{B}$
11	$A - 1 = A + \text{All 1's} + 0$	A	1

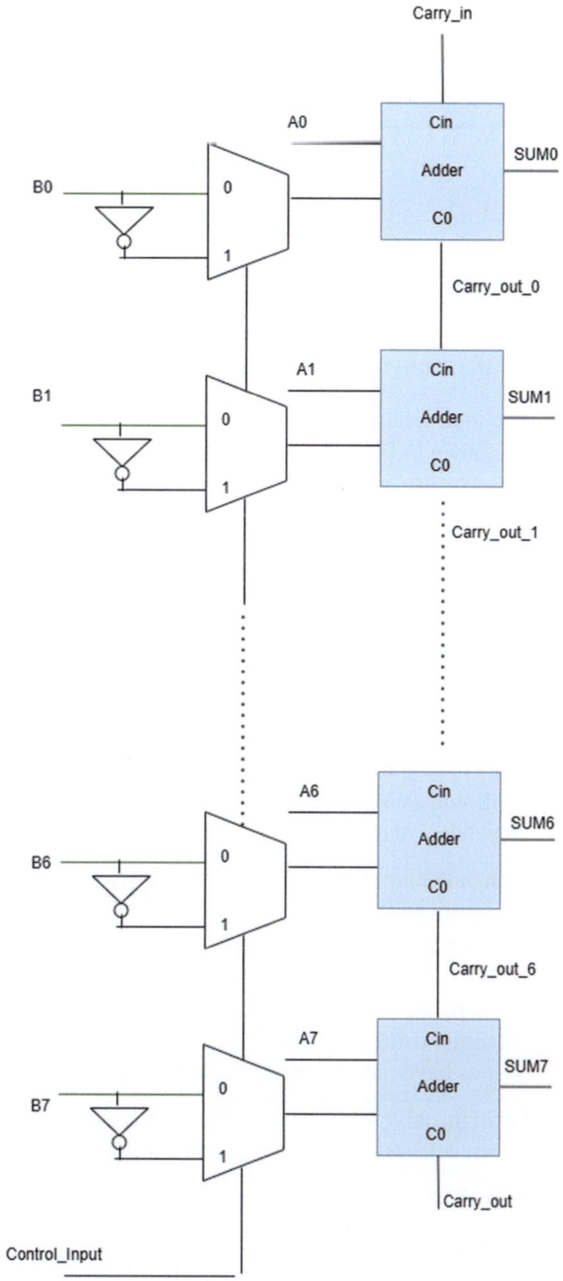

Fig. 4.4 Logic design of 8-bit adder, subtractor

Table 4.3 Operational table of arithmetic unit

Opcode S1 S0	Operation
00	A
01	A + B
10	A − B − 1
11	A − 1

1. From the table it is clear that the X input of adder is always A and Y input of adder is 0, B, $\overline{\boldsymbol{B}}$, 1 depending on the status of the opcode.
2. So, we need to have 4:1 multiplexer to pass one of the values 0, B, $\overline{\boldsymbol{B}}$, 1 depending on the status of the opcode (Table 4.3).

The logic design of the 8-bit arithmetic unit is shown in Fig. 4.5.

As shown in Fig. 4.5, the design uses eight 4 to 1 multiplexers, 8 two to 1 multiplexers to pass the 0, B, $\overline{\boldsymbol{B}}$, 1 for the opcode = 00,01,10,11 respectively to the Y input of the adder. The design uses the 8 full-adders. Here opcode is 2-bit and indicated by S1, S0 lines.

4.2.1 *Design Optimization Using Least Resources*

Now most of the times we stop here only and we never think to optimize the design further. Now let us ask ourself that is there any chance for the optimization? Answer you will find as Yes.

Now how we can optimize. Use the basic fundamentals of the Boolean algebra.

1. When the inputs of 4:1 mux is connected to logic 1, logic 0 then the design can be optimized.
2. We can use the 2:1 mux instead of the 4:1 mux and can reduce the area. We will have least gate count design.
3. Now focus on the Y input of adder which is generated by the 4:1 mux and let us design it using the 2:1 mux.
4. Let us have the optimization Table 4.4.

From the table it is clear that for the S1S0 = 01 the Y is B and for the S1S0 = 10 Y is complement of B so we have two product terms **S0.*B*, S1.$\overline{\boldsymbol{B}}$**

That is SOP equation is $\boldsymbol{Y = S0.B + \mathrm{S1}.\overline{B}}$

Now let us validate the truth-table to check for the correctness of the data at Y input of adder (Table 4.5).

Now we can conclude that,

$$\mathrm{Y} = \mathrm{S0.B} + \mathrm{S1}.\overline{\mathrm{B}}.$$

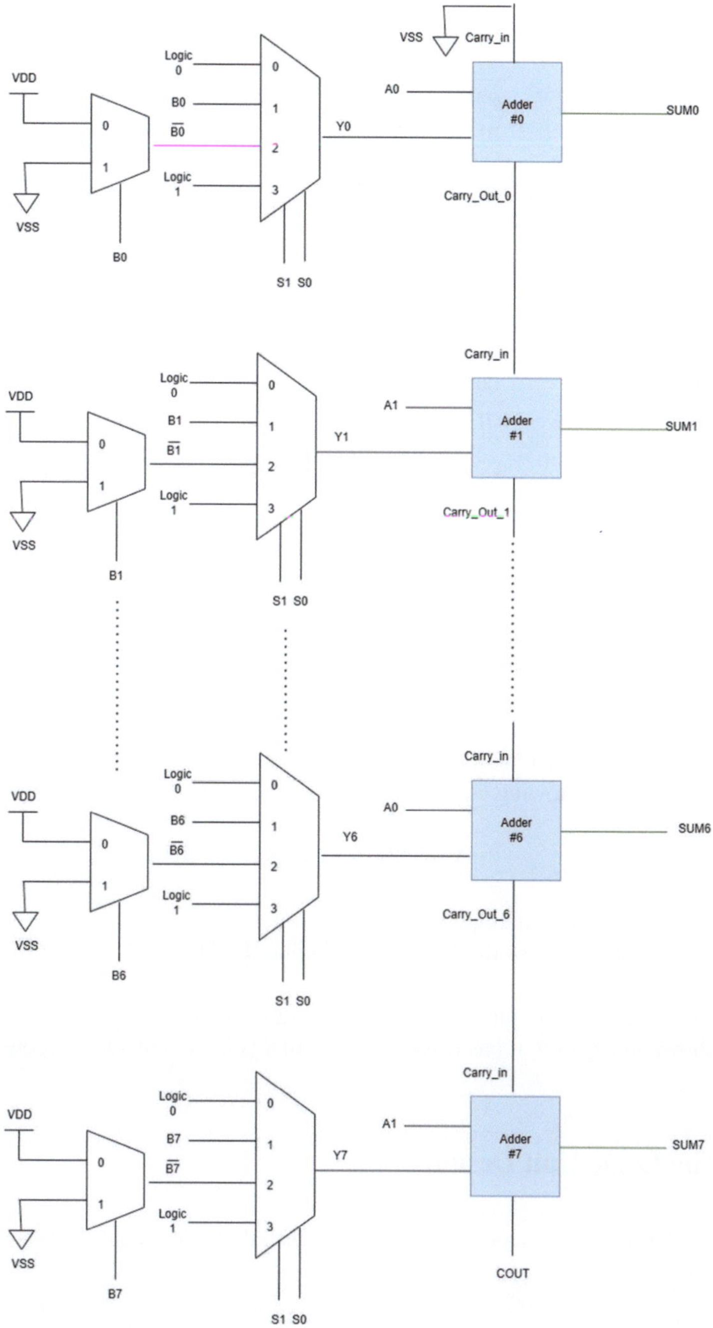

Fig. 4.5 8-Bit arithmetic unit design

Table 4.4 Optimization table for the Y input entries

Opcode S1 S0	Y input of adder
00	0
01	B
10	$\overline{B}$
11	1

Table 4.5 Table of the operations of arithmetic unit

Opcode S1 S0	$S0.B$	$S1.\overline{B}$	Y input of adder $Y = S0.B + S1.\overline{B}$
00	0	0	$0+0$
01	B	0	$0+\text{B}$
10	0	$\overline{B}$	$0+\overline{B}$
11	B	$\overline{B}$	$B+\overline{B}=1$

Table 4.6 Optimized table of 2:1 mux to have Y

B	$Y = S0.B + S1.\overline{B}$
0	S1
1	S0

The entries and functionality is correct and we can optimize the 4:1 mux using the 2:1 mux as the following equation is 2:1 mux equation

$$\text{Y} = \text{S0.B} + \text{S1.}\overline{\text{B}}.$$

So, what are inputs and outputs of 2:1 mux (Table 4.6).

As shown in Fig. 4.6 for the B = 0 the Y input of adder is S1 and for the B = 1 Y input of adder is S0.

The design uses the eight 2:1 multiplexer, eight adders and as compared to the design shown in Fig. 4.5, it has least resources that is lesser area, lesser power.

4.3 The Logic Unit Design

Now consider the design requirement is to perform the following operations on two 8-bit numbers (Fig. 4.7).

1. A OR B.
2. A XOR B.
3. A AND B.
4. NOT A.

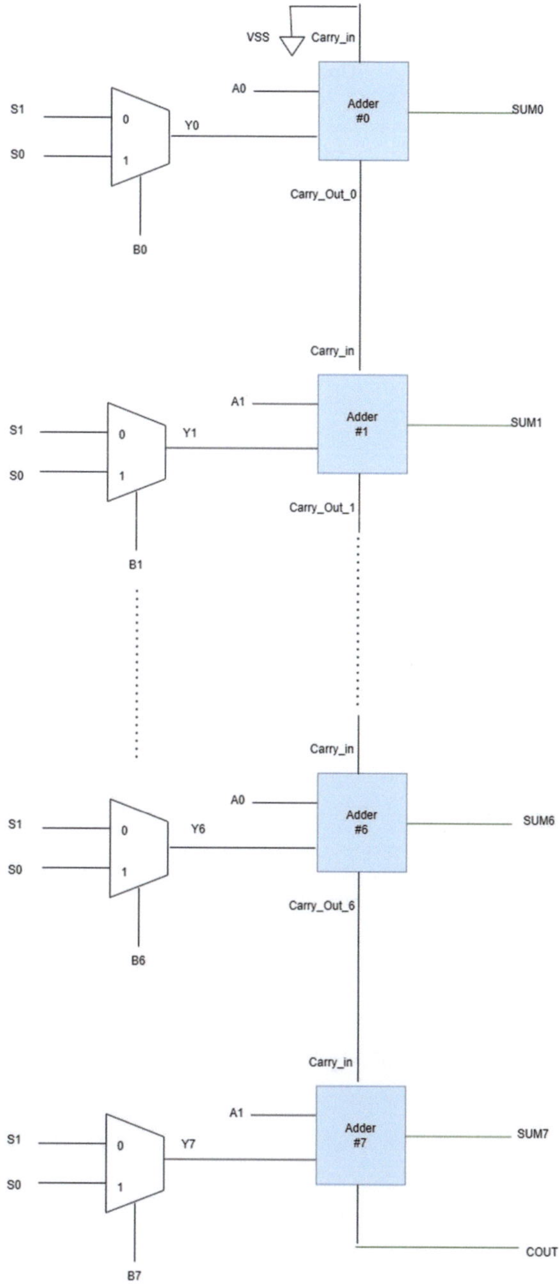

Fig. 4.6 Optimized arithmetic unit

Opcode S1 S0	Operation
00	A OR B
01	A XOR B
10	**A AND B**
11	NOT A

Fig. 4.7 Logic unit functional table

Opcode S1 S0	Operation
00	A OR B
01	A XOR B
10	A AND B
11	NOT A

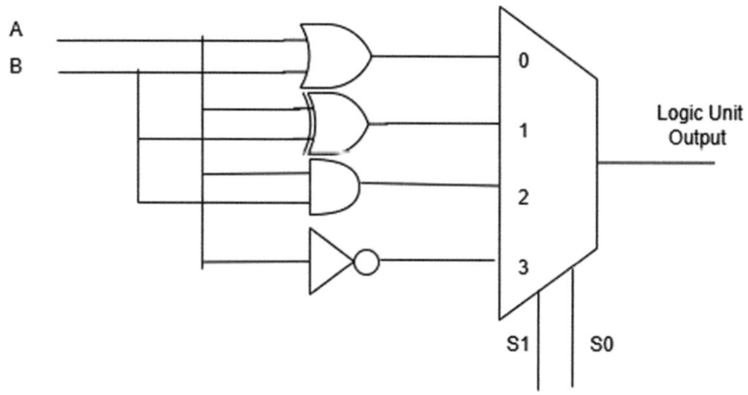

Fig. 4.8 Logic unit design

What we need the 4:1 mux and the logic gates to perform one of the operations at a time depending on the S1, S0 status that is control code or opcode for the logic unit. For four operation we have 2-bit opcode.

The design of 1-bit logic unit is shown in Fig. 4.8.

Now as we need to have 8-bit logic unit we need eight OR, XOR, AND, NOT gates and eight 4:1 multiplexer. So huge resources and we should discuss the strategy to optimize the area.

4.4 The ALU Design

Consider the four arithmetic and four logic operations.

1. Parallel input parallel output of A.
2. Addition of A, B.
3. Subtraction of A, B, 1.
4. Decrement A.
5. A OR B.
6. A XOR B.
7. A AND B.
8. NOT A.

We need to have 3-bit opcode (s2, s1, s0) as we have 8 operations to perform. The design should perform the single operation at a time either logical operation or arithmetic operation. Let us perform the logical operation when MSB of opcode that is $S2 = 1$ and let us perform the arithmetic operation for the $S2 = 0$. The truth-table is shown in Table 4.7.

Table 4.7 Truth-table of the logical unit

Opcode S2 S1 S0	Operation
000	A
001	A + B
010	A − B − 1
011	A − 1
100	A OR B
101	A XOR B
110	A AND B
111	NOT A

What we need is

1. The 2:1 mux at the output to get the result of the arithmetic unit or logic unit. Refer the above section for the details of the arithmetic unit, logic unit design.
2. We need logic unit and arithmetic unit.
3. For 8-bit operands we need to have following resources
   a. Eight 2:1 mux.
   b. Eight instances of the arithmetic unit.
   c. Eight instances of the logic unit.

The design is shown in Fig. 4.9.

The issues in the above design

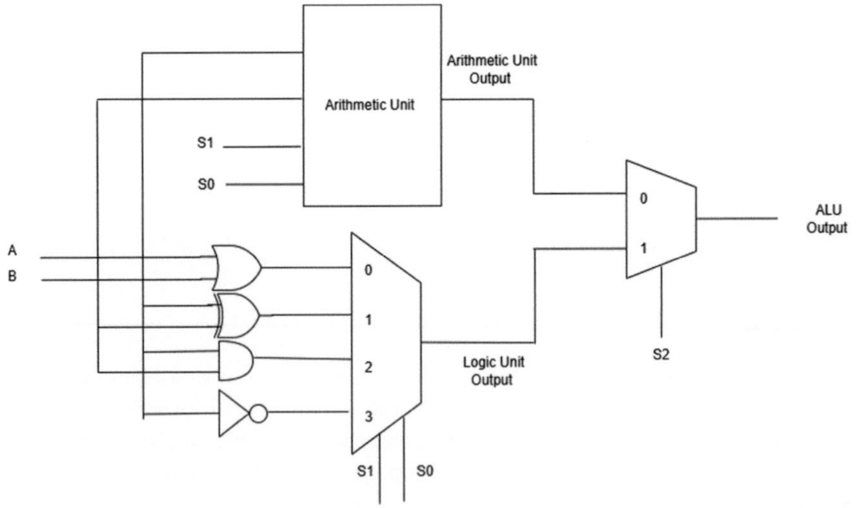

Fig. 4.9 Inefficient ALU design

1. Performs both the arithmetic and logic operations at a time.
2. More area, more power.
3. Separate logic for the arithmetic unit and logic unit.
4. Multiplexers at the output.

So, as discussed in the Chap. 3 let us optimize the design. Let us have area optimization.

4.5 The Efficient ALU Design: Area Optimization

Now consider the design optimization for the following eight operations.

1. Parallel input parallel output of A.
2. Addition of A, B.
3. Subtraction of A, B, 1.
4. Decrement A.
5. A OR B.
6. A XOR B.
7. A AND B.
8. NOT A.

What we need to do is following

1. Let us perform the arithmetic and logical operations using adder. Why? Because for the S1S0 as 00, 01, 10, 11 the Y input of adder is 0, B, complement of B, logic 1 for the arithmetic and logical operations.
2. Let us have common resource as a adder. To perform the 8-bit operations, we need to have the 8 full-adders.
3. Let us include the details in the table that is adder inputs

Opcode S2 S1 S0	Operation	X input of adder	Y input of adder
000	A = A + 0 + 0	A	0
001	A + B = A + B + 0	A	B
010	$A - B - 1 = A + \overline{B} + 1 - 1 = A + \overline{B} + 0$	A	$\overline{B}$
011	A − 1 = A + All 1's + 0	A	1
100	A = (A OR B) + 0 + 0	A OR B	0
101	A XOR B = A XOR B + 0	A	B
110	A AND B = A.B XOR $\overline{B}$	A AND B	$\overline{B}$
111	NOT A = A + All 1's + 0	A	1

4. From the table, it is clear that the X input of adder is always A for the arithmetic operation and for the logical operation we need to disable the carry propagation that is carry_out from each adder should be zero then we can use adder as a resource to perform the logical operations.
i. So, when S2 = 1 then let us disable the carry propagation using the logic.
j. For S1 = 0, S0 = 0 the X input should be A OR B so have the product term $\overline{S1}$. $\overline{S0}$. B.
k. For S1 = 0, S0 = 1 the X input should be A.
l. For S1 = 1, S0 = 0 the X input should be A OR B so have the product term $\overline{S1}$. $\overline{S0}$. B.
m. For S1 = 0, S0 = 0 the X input should be A OR complement of B so have the product term $S1$. $\overline{S0}$. $\overline{\boldsymbol{B}}$.

The design to generate the required input at X is shown in Fig. 4.10.

5. For the S2 = 0 that is arithmetic operation X is always A as required.
6. The Y input of adder is 0, B, $\overline{\boldsymbol{B}}$, 1 depending on the status of the opcode.
7. We will use the optimize logic of the arithmetic unit and at X input the combo logic.

The logic design of the 8-bit arithmetic logic unit is shown in Fig. 4.5.
As shown in the design the design uses least resources (Fig. 4.11)

1. Only 8 full-adders to perform both the arithmetic and logical operations
2. Only eight 2:1 mux
3. Only 8 combo logic

From the design it is clear that it is area optimized design and performs only one operation at a time.

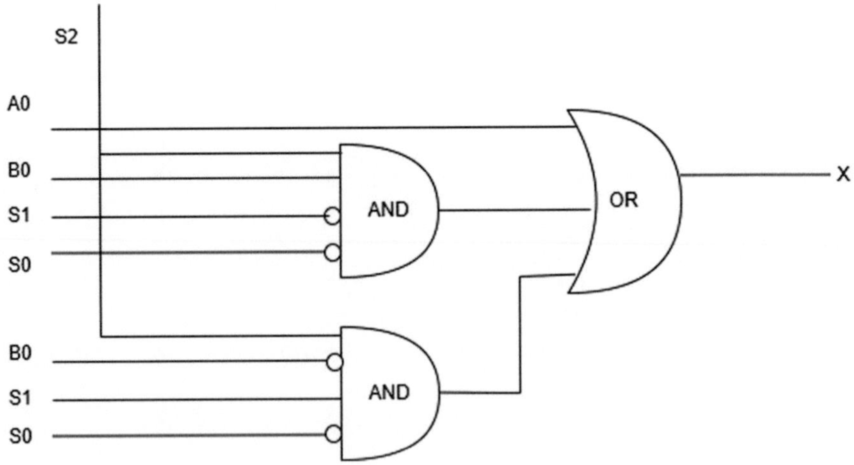

Fig. 4.10 Combo logic

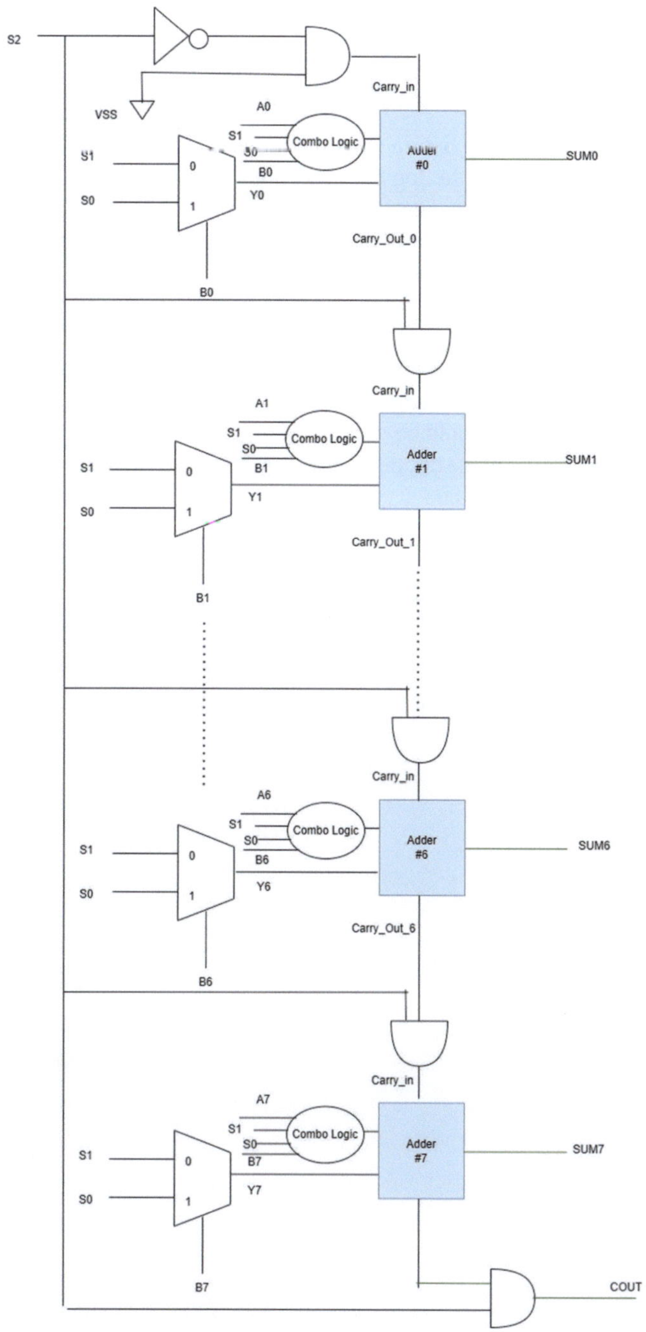

Fig. 4.11 Optimized design of the 8-bit ALU

4.6 Important Takeaways

Following are few of the important points to summarize this chapter.

1. The arithmetic resource sharing should be used to optimize the area
2. The adder can perform the logical operations if the A, B inputs are managed properly with the logic elements.
3. To perform the 8-bit XOR using ripple carry adder disable the carry out of each adder that is mask carry output
4. To perform the 8-bit NOT using the adder pass A, all 1's and stop carry propagation.
5. Don't use the chain of multiplexers at output to select the output of the arithmetic unit or logical unit as it consumes more area, power.
6. The resource sharing should be used to optimize the area of the ALU
7. The logic replication of the duplication should be used to have the same functional logic at the inputs of the adder.

4.7 Puzzle

Design the ALU to perform the operations PIPO, Addition, subtraction, add with carry, sub with borrow, increment A, decrement A, OR, XOR, AND, NOT? Can you optimize the logic unit using minimum logic elements.

Chapter 5
Practical Scenarios: Area Optimization

In the previous few chapters, we have discussed the combinational design elements. By considering the area requirement for any kind of the design we need to go for the optimization phase. The chapter covers the practical scenarios, area optimization techniques, logic duplications, cascade versus parallel logic, delay optimization and area optimized designs.

5.1 Cascade Mux 4:1

As discussed in the previous few chapters the multiplexers are widely used as data selectors. We have the priority logic and parallel logic. The multiplexers can be designed to have the cascade stages of the multiplexers and useful as to process the slow arriving inputs and fast arriving inputs.

Effectively the goal of the designer is to balance the delay based on the arrival of the inputs in the data path.

Consider the 4:1 mux design using the minimum number of 2:1 mux. As shown in the Figure the design uses three cascade 2:1 mux and decoding logic. Table 5.1 The 4:1 mux truth table describes the truth table of 4:1 mux.

The multiplexer shown in Fig. 5.1 The priority mux is priority logic and has three cascade multiplexers. The disadvantage is it needs the decoding logic and has additional area. Here the I0 has the highest priority and I3 has the lowest priority.

The generation of D2, D1 Do is shown in the able 2 The decoding logic of the priority mux (Table 5.2).

V. Taraate, *Advanced Digital Design Techniques*,
https://doi.org/10.1007/978-981-95-4510-0_5

Table 5.1 The 4:1 mux truth table

S1	S0	Y
0	0	I0
0	1	I1
1	0	I2
1	1	I3

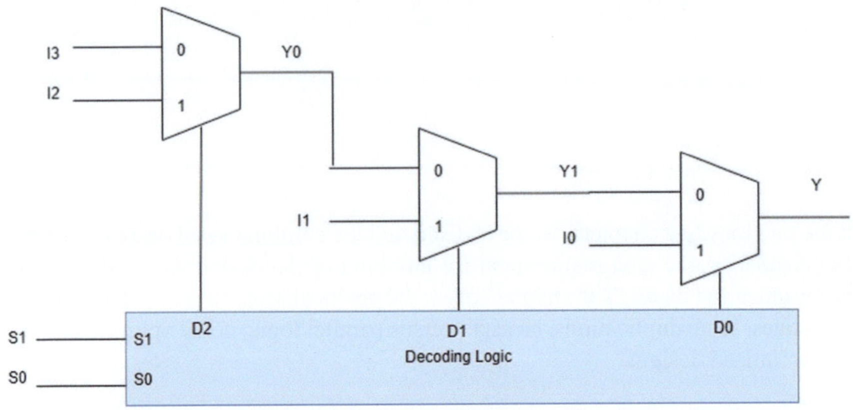

Fig. 5.1 The priority mux

Table 5.2 The decoding logic of the priority mux

S1	S0	D2	D1	D0	Y
0	0	0	0	1	I0
0	1	0	1	0	I1
1	0	1	0	0	I2
1	1	0	0	0	I3

5.2 Propagation Delay

Consider the priority mux design, the 4:1 mux is designed using the three 2:1 mux The main issue in the design is longest path for the I3 as I3 has the lowest priority. If I3 is late arriving signal and I0 is fast arriving signal then the Y = I0 for S1 = 0 and S2 = 0 and has least propagation delay. But if I3 is fast arriving signal and S1 = 1 and S2 = 1 then after 3 * tpd the multiplexer output Y = I3. Where tpd is propagation delay of the 2:1 mux (Fig. 5.2).

If we consider the tpd of each mux is 0.5 ns then the total delay is 3 * 0.5 ns = 1.5 ns.

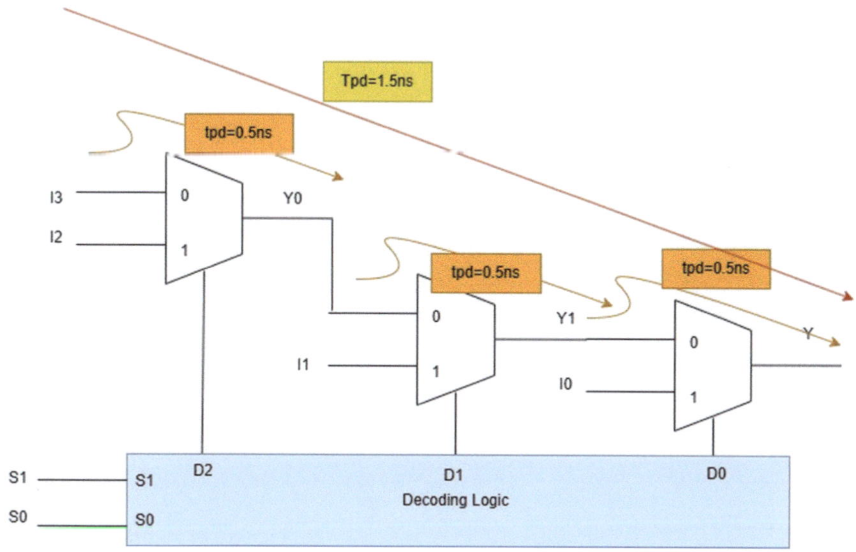

Fig. 5.2 The propagation delay of the priority mux

5.2.1 *Input Swapping: Late Versus Faster Arrival Signals*

Now consider that the design has faster arrival signal I3 and slower arrival signal I0. To balance the delay, we can swap the inputs I3, I0 as shown in Fig. 5.3 Input I3, I0 swapping to balance the delay.

If the I3 is arrival time is 2 ns and I0 arrival time is 0 ns then the decoding logic should be such that the late arrival signal I3 should not incur the more delay. Hence I3 is used as one of the inputs of 2:1 mux at the output stage and I0 is pushed last at the input stage 2:1 mux.

5.3 Hazards

Due to the variable path delays there are fluctuations in the outputs and circuits are prone to the hazards. The unwanted switching transients due to the finite propagation delays are called as hazards.

There are three different kinds of hazards in the digital circuits

1. Static hazard
2. Dynamic hazard
3. Functional hazard.

The static hazards are due to changes in the input which causes the changes in an output momentarily before stabilizing to the valid value.

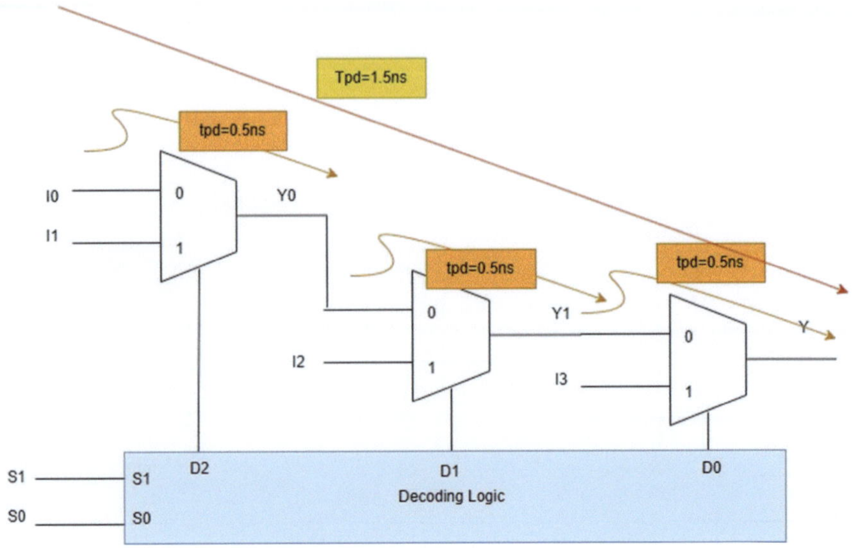

Fig. 5.3 Input I3, I0 swapping to balance the delay

- **Static-1 Hazard:** Static-1 Hazards occur in SOP (SUM-OF-PRODUCT) circuit.
  1. If the output is currently at logic level 1 and after the input changes its state, the output momentarily changes to 0 before settling to logic 1, then it is a Static-1 hazard.
- **Static-0 Hazard:** Static-0 Hazards occur in the POS (PRODUCT-OF-SUM) circuit.
  2. If the output is currently at logic state 0 and after the input changes its state, the output momentarily changes to 1 before settling to logic 0, then it is a Static-0 hazard (Fig. 5.4).

***Let us consider static-1 hazard*.**

To detect a static-1 hazard for a digital circuit following steps are used:

- **Step-1**: Write down the output of the digital circuit, say **Y**.

Fig. 5.4 The static hazards

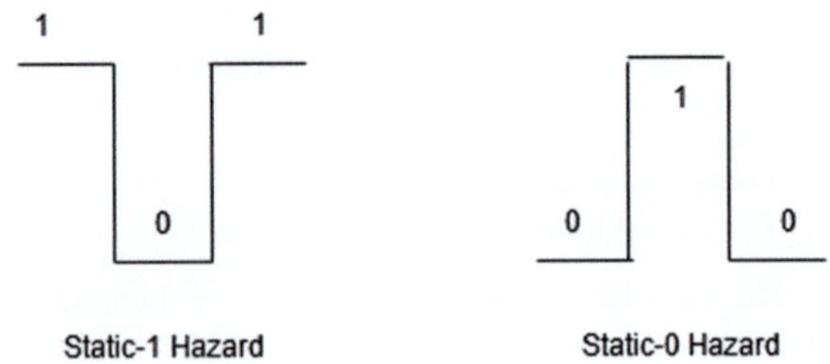

- **Step-2**: Draw the K-map for this function **Y**
- **Step-3**: Check for the prime implicates and the pair of 1's which causes the static-1 hazard

Let us consider the 2:1 mux shown below (Figs. 5.5, 5.6; Table 5.3).

The truth table is shown below

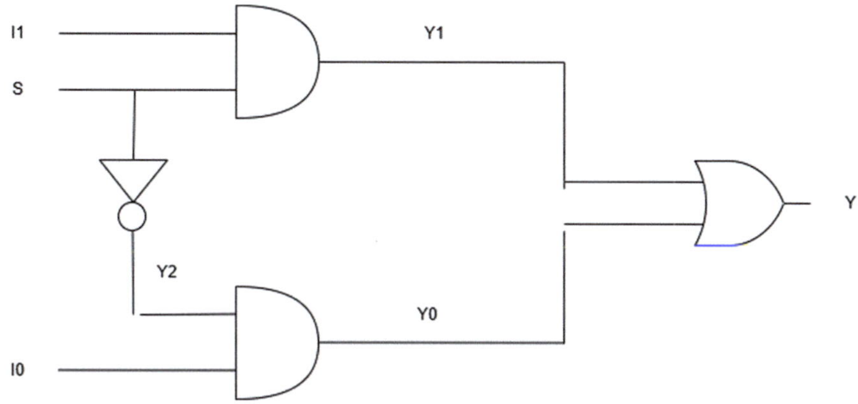

Fig. 5.5 The 2:1 mux logic

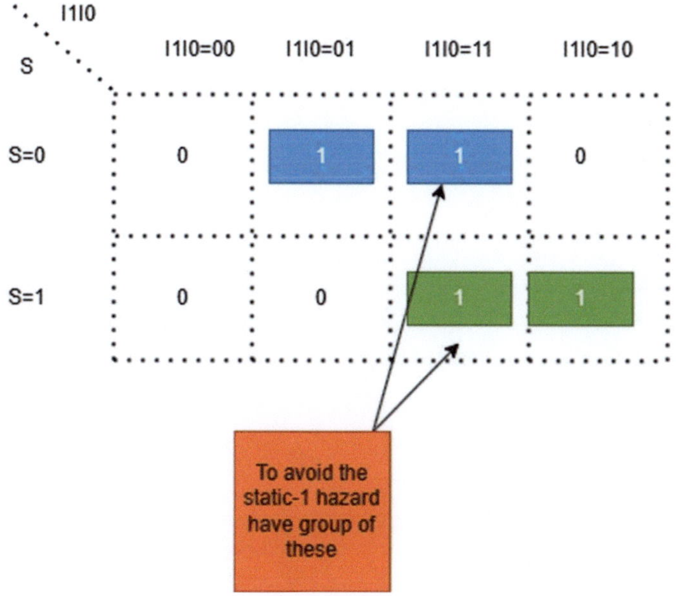

Fig. 5.6 The grouping of pair of 1's to remove the static-1 hazard

Table 5.3 The truth table of the 2:1 mux

S	I1	I0	Y
0	0	0	0
0	0	1	1
0	1	0	0
0	1	1	1
1	0	0	0
1	0	1	0
1	1	0	1
1	1	1	1

We have output, say F, as: F(S, I1, I0) = SI1 + ~ SI0 = ∑m{1,3,6,7} F(S,I1,I0)} Lets sketch the K-map for this (Table 5.3).

The logic expression is given by Y= S.I1 + ~S. I0 + I1.I0

To remove the static-1 hazard let us include the terms by including the logic gate (Fig. 5.7).

Similarly, for Static-0 Hazards, we need to consider logic 0's instead of 1's, and use the POS expression instead of the SOP and group adjacent 0's to avoid the static 0 hazards.

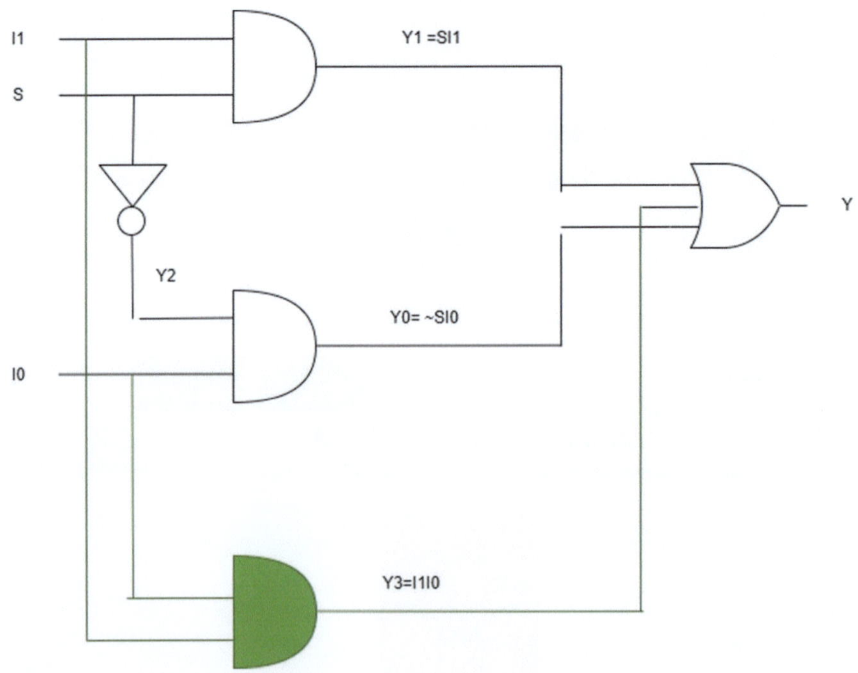

Fig. 5.7 The static-1 hazard removal circuit

Even if we design the RISC-V pipelined stages then we need to work on the hazard removal logic. We can have three types of the hazards.

1. Data hazards
2. Structural hazards
3. Control Hazards.

More details on the pipelined processors please refer Chap. 13.

5.4 Glitches

If the combinational logic path delays are imbalanced then the glitches occurs in the design.

The glitches in the design are shown in the Fig. 5.8.

As shown in Fig. 5.9 The glitches in the design, if we consider the propagation delay then the design is prone to glitches and this is due to the buffer delay in the path B. This is the imbalanced combo delay and these kinds of paths are prone to the glitches.

To have the glitch free designs do the following

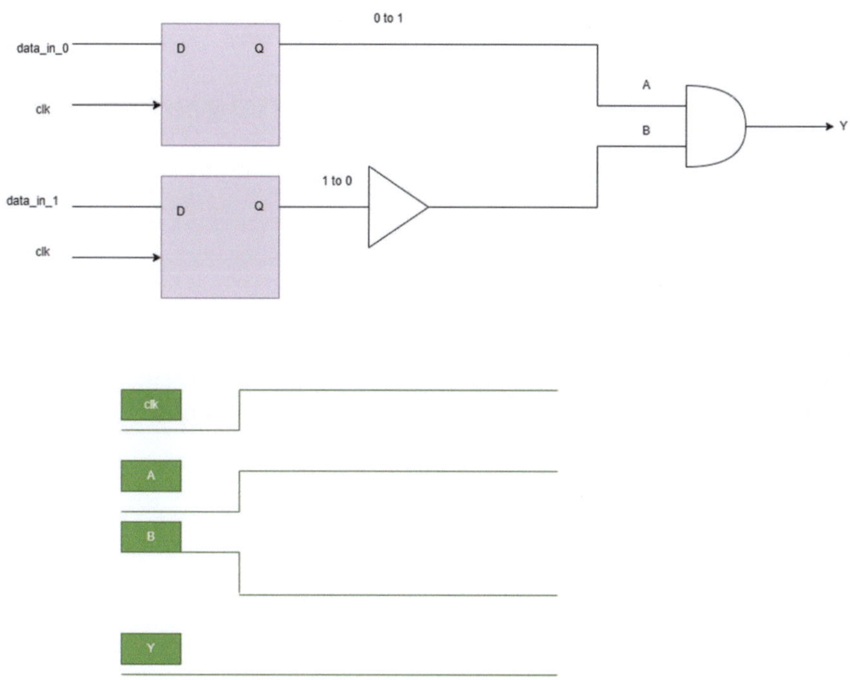

Fig. 5.8 The zero delay waveforms

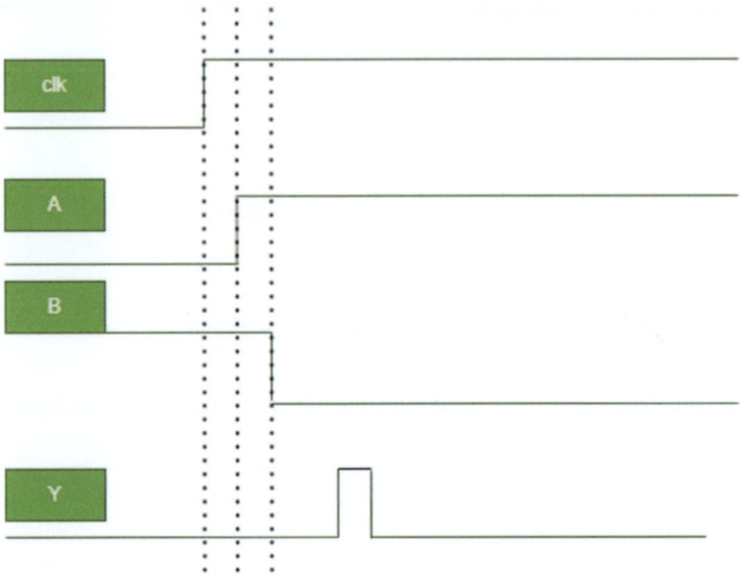

Fig. 5.9 The glitches in the design

1. Identify which nets have glitches and how many data paths are glitch free and how many have the glitch behaviour
2. Check for the glitch duration and power consumptions
3. Remove glitches and recover power use the re-sizing and buffer insertions to do the same.

5.5 Decoding of the Data Input

For the selection of one of the memories or IO devices in the design we need to have the decoding logic, The truth table of the 2:4 Decoder having active high output is shown in Table 5.4. The 2:4 active high output decoder truth table below. The decoder has the active high enable and two select lines and four output lines.

Table 5.4 The 2:4 active high output decoder truth table

en	*s*1	*s*0	*y*3	*y*2	*y*1	*y*0
1	0	0	0	0	0	1
1	0	1	0	0	1	0
1	1	0	0	1	0	0
1	1	1	1	0	0	0
0	x	x	0	0	0	0

For the *en* = 1 the decoder is enabled and when *en* = 0 the decoder outputs are all zero. The relationship between the input lines $n = 2$ and output lines $m = 4$ is given by $m = 2^n$.

The Boolean equations for the outputs y0 to y3 are given below and the logic diagram of the 2:4 decoder is shown in Fig. 5.10 The logic of the 2:4 decoder. If the memory of IO device has the active high chip select then the decoder having the active high outputs are used.

$$y0 = en.\overline{s1}.\overline{s0}$$

$$y1 = en.\overline{s1}.s0$$

$$y2 = en.s1.\overline{s0}$$

$$y3 = en.s1.s0$$

Consider the memories or IO devices which has the active low chip select. To select the one of the IO or memory devices we can use the decoders having the active low chip-select. The truth table of the 2:4 decoder having active low outputs is shown in Table 5.5. The 2:4 decoder active low output truth table. As shown one of the outputs of the decoder is active low at a time and it is function of the s1, s0. For the en = 1 the decoder is enabled and for the en = 0 the decoder outputs are all 1's.

The 2:4 decoder active low output (Table 5.5).

Fig. 5.10 The logic of the 2:4 decoder

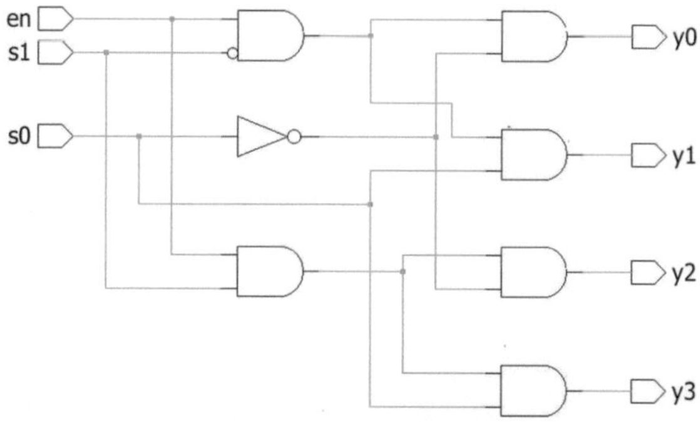

Table 5.5 The 2:4 decoder active low output truth table

en	*s*1	*s*0	*y*3	*y*2	*y*1	*y*0
1	0	0	1	1	1	0
1	0	1	1	1	0	1
1	1	0	1	0	1	1
1	1	1	0	1	1	1
0	x	x	1	1	1	1

Table 5.6 The 4:16 decoder truth table

en	s3	s2	s1	s0	y15	y14	y13	y12	y11	y10	y9	y8	y7	y6	y5	y4	y3	y2	y1	y0
1	0	0	0	0	0	0	0	0	0	0	0	0	0	0	0	0	0	0	0	1
1	0	0	0	1	0	0	0	0	0	0	0	0	0	0	0	0	0	0	1	0
1	0	0	1	0	0	0	0	0	0	0	0	0	0	0	0	0	0	1	0	0
1	0	0	1	1	0	0	0	0	0	0	0	0	0	0	0	0	1	0	0	0
1	0	1	0	0	0	0	0	0	0	0	0	0	0	0	0	1	0	0	0	0
1	0	1	0	1	0	0	0	0	0	0	0	0	0	0	1	0	0	0	0	0
1	0	1	1	0	0	0	0	0	0	0	0	0	0	1	0	0	0	0	0	0
1	0	1	1	1	0	0	0	0	0	0	0	0	1	0	0	0	0	0	0	0
1	1	0	0	0	0	0	0	0	0	0	0	1	0	0	0	0	0	0	0	0
1	1	0	0	1	0	0	0	0	0	0	1	0	0	0	0	0	0	0	0	0
1	1	0	1	0	0	0	0	0	0	1	0	0	0	0	0	0	0	0	0	0

(continued)

Table 5.6 (continued)

1	1	0	1	1	0	0	0	0	1	0	0	0	0	0	0	0	0	0	0	0
1	1	1	0	0	0	0	0	1	0	0	0	0	0	0	0	0	0	0	0	0
1	1	1	0	1	1	0	1	0	0	0	0	0	0	0	0	0	0	0	0	0
1	1	1	1	0	0	1	0	0	0	0	0	0	0	0	0	0	0	0	0	0
1	1	1	1	1	1	0	0	0	0	0	0	0	0	0	0	0	0	0	0	0
0	x	x	x	x	0	0	0	0	0	0	0	0	0	0	0	0	0	0	0	0

The Boolean expressions of the 2:4 decoder having active low outputs are shown in below. The logic circuit of the 2:4 decoder having active low outputs is shown in Fig. 5.11 The logic diagram of the 2:4 decoder having active low output

$$y0 = \overline{en.\overline{s1}.\overline{s0}}$$

$$y1 = \overline{en.\overline{s1}.s0}$$

$$y2 = \overline{en.s1.\overline{s0}}$$

$$y3 = \overline{en.s1.s0}$$

The decoder 4:16 is realized using the five 2:4 decoders. Each decoder has the active high outputs and active high enable.

The truth table of the 4:16 decoder is shown below (Table 5.6).

As shown on the table, we have four groups depending on the status of $s3$, $s2$. The $s3\ s2 = 00,01,10,11$ and hence we need to have the main decoder or master decoder to enable one of the decoders. The four entries for the $s3s2 = 00$ needs decoder_1, for the $s3s2 = 01$ needs decoder_2, for the $s3s2 = 10$ needs decoder_3 and for the $s3s2 = 11$ needs decoder_4.

The design of the 4:16 decoder using the minimum number of the 2:4 decoders is shown in Fig. 5.12 The 4;16 decoder using 2:4 decoder.

Fig. 5.11 The logic diagram of the 2:4 decoder having active low output

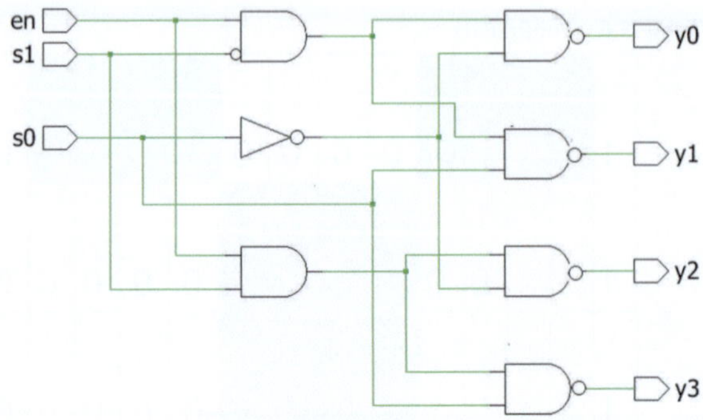

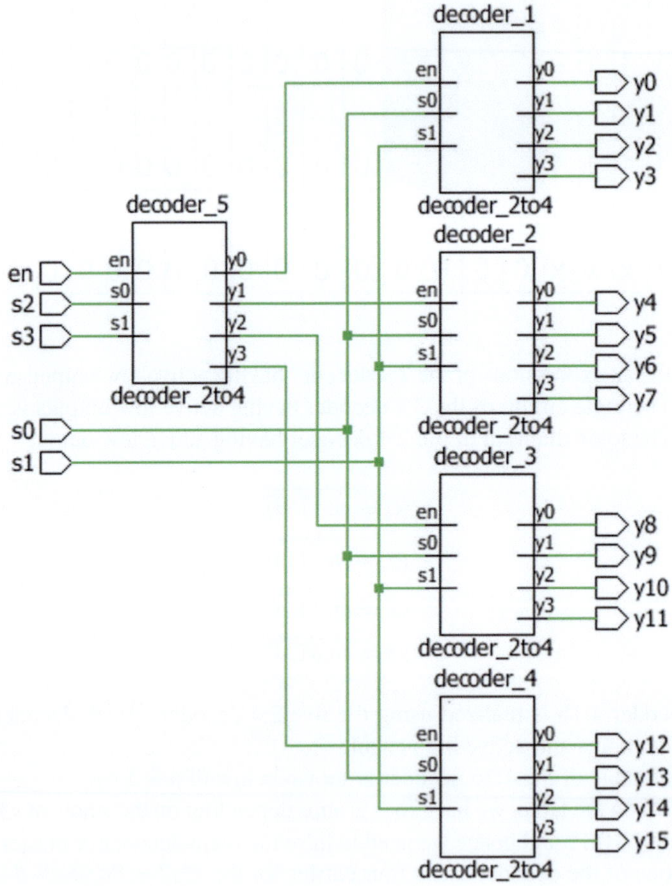

Fig. 5.12 The 4;16 decoder using 2:4 decoder

As shown in Fig. 5.12 The 4;16 decoder using 2:4 decoder, for the $s3s2 = 00$ the decoer_1 is enabled, for the $s3s2 = 01$ the decoder_2 is enabled, for the $s3s2 = 10$ the decoder_3 is enabled and for the $s3s2 = 11$ the decoder_4 is enabled.

5.6 Encoding of the Data Input

Let us understand the encoding of the data inputs. We need to have the encoder which performs the reverse operation of the decoder. Consider the four input lines I3 to I0 and receives the level sensitive input signals. The design needs to generate the binary equivalent of the input. That is for the I3 it should generate output as 11 and for the I0 it should generate output as 00. For such requirement we need to have the 4:2 encoder. The truth table of the 4:2 encoder is shown below (Table 5.7 The 4:2 encoder truth table).

The Boolean equations of $y0$, $y1$ are given below (Fig. 5.13)

$$y1 = I3 + I2$$

$$y0 = I3 + I1$$

As shown in the Fig. 5.7 the issue in the design of the 4:2 encoder is, the $I0$ is unconnected and for the $I3 = 0$, $I2 = 0$, $I1 = 0$ and $I0 = 0$ also the output of the encoder is 00 but this is invalid output as all the inputs are zero. These issues can be resolved by generating an output "invalid_output" and shown in Table 5.8 The practical 4:2 encoder truth table.

Table 5.7 The 4:2 encoder truth table

I3	I2	I1	I0	y1	y0
1	0	0	0	1	1
0	1	0	0	1	0
0	0	1	0	0	1
0	0	0	1	0	0

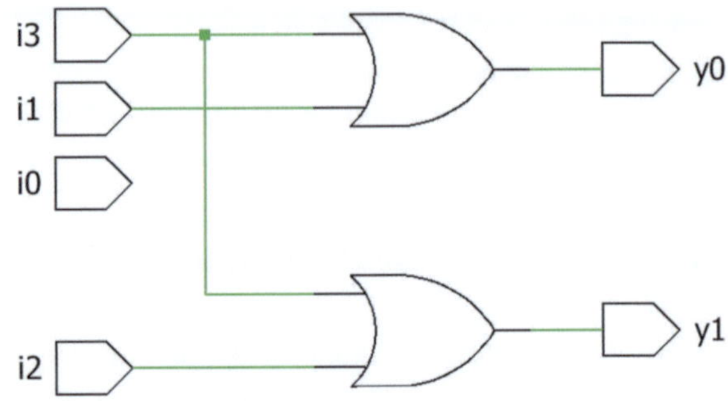

Fig. 5.13 The logic structure of the 4:2 encoder

Table 5.8 The practical 4:2 encoder truth table

I3	I2	I1	I0	y1	y0	Invalid_output
1	0	0	0	1	1	0
0	1	0	0	1	0	0
0	0	1	0	0	1	0
0	0	0	1	0	0	0
0	0	0	0	0	0	1

If all the inputs are 0's then the encoder outputs are 00 and invalid_output=1 and indicates that ignore this output. The design is shown in Fig. 5.14 The logic structure of the 4:2 encoder with invalid output flag

Most of the time we need to have the priority encoders in the design. Consider the design of the priority logic where I3 input has highest priority and $I0$ has lowest priority. In such scenario when $I3 = 1$ then all other inputs need to be considered as don't care(x) and accordingly the logic needs to be derived. When the $I3 = 0$ then only check for the $I2$ input status. This logic is called as priority encoding logic and described in Table 5.9 The 4:2 priority encoder truth table.

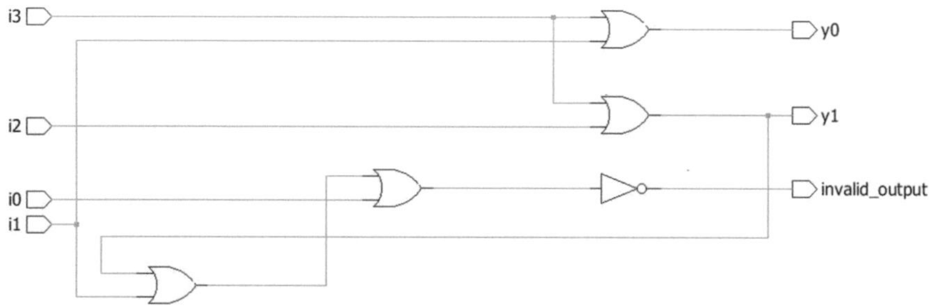

Fig. 5.14 The logic structure of the 4:2 encoder with invalid output flag

Table 5.9 The 4; 2 priority encoder truth table

*I*3	*I*2	*I*1	*I*0	*y*1	*y*0	Invalid_output
1	x	x	x	1	1	0
0	1	x	x	1	0	0
0	0	1	x	0	1	0
0	0	0	1	0	0	0
0	0	0	0	0	0	1

The design is shown in Fig. 5.15 The logic structure of the 4:2 priority encoder below. The Boolean equations are listed below.

$$y1 = I3 + I2$$
$$y0 = I3 + \overline{I2}.I1$$
$$\text{invalid\_output} = \overline{I3 + I2 + I1 + I0}$$

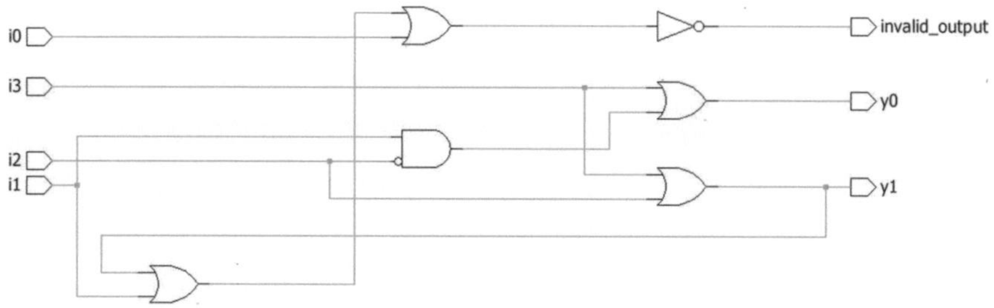

Fig. 5.15 The logic structure of the 4:2 priority encoder

5.7 Logic Duplication

Consider the design discussed in the above section, that is the design of the 4:16 decoder using the minimum number of decoders. We need to have the five 2:4 decoders to design the 4:16 decoder. If we have the FPGA design having the four inputs look up tables then the design needs 4 * 1 + 16 * 1 = 20 LUTs.

But for the larger decoders like 8:256 decoder the design needs to have the 8 input LUTs and the FPGA LUT count is almost around 256 * 3 = 768 LUTs at output side and 16 * 2 = 32 LUTs at input side, Total LUTs 768 + 32 = 800.

If we consider the following structure of the CLB having the LUTs and use the logic duplication using the AND arrays then we can have the 256 LUTs for the AND array and for the two 4:16 decoders we can have the 32 LUTs. So, the optimization is 800–256-32 = 816–288 = 528LUTs thus for the FPGA based designs the logic optimization is used to reduce the area (Fig. 5.16).

Note: The logic optimization reduces the area for the larger density encoders and decoders and not useful for the smaller designs.

The logic duplication for the 4:16 decoder is shown below (Fig. 5.17 The logic duplication) and has two 2:4 decoders and 16 AND gates at the output.

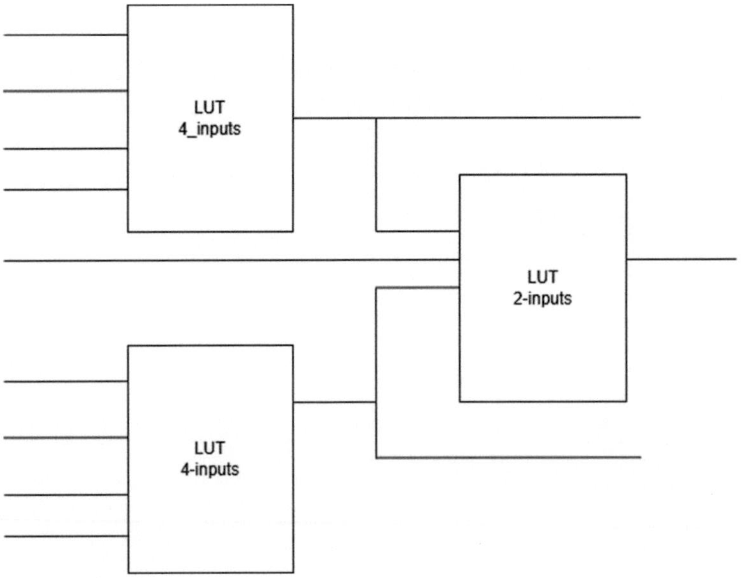

Fig. 5.16 The custom LUT structure

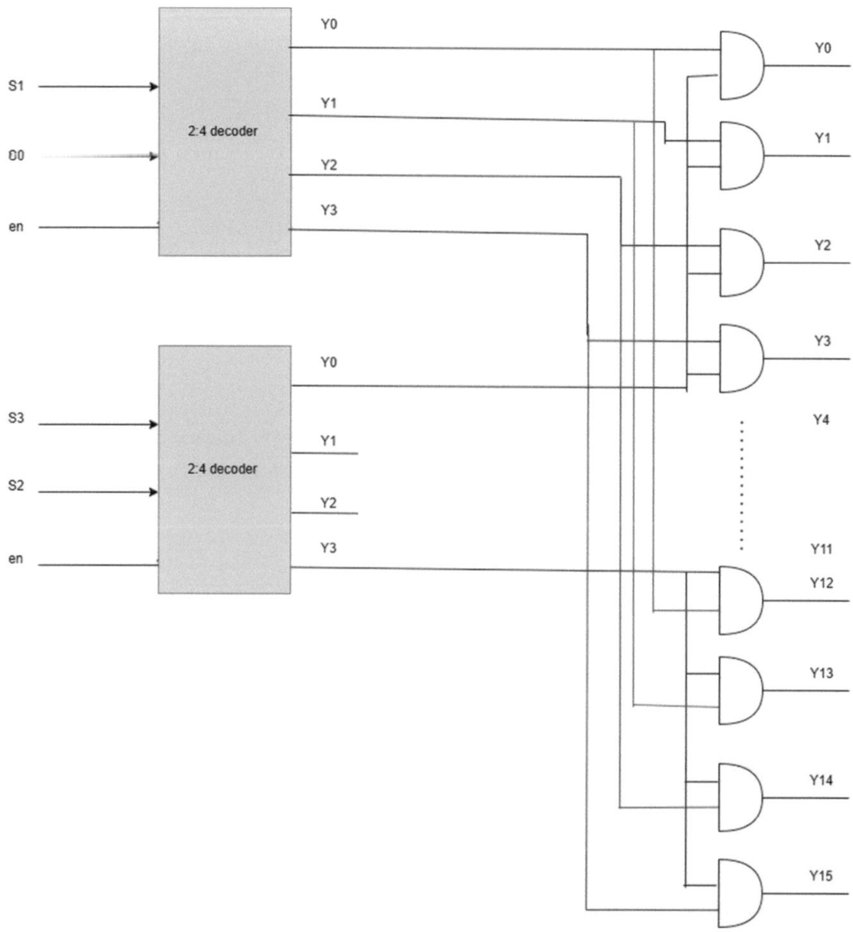

Fig. 5.17 The logic duplication

5.8 Grouping

As discussedin the previous chapters, we need to have the digital design having least resources. Consider we have 4-bit number and needs to have the parity checker at the output; we can design this by using cascade XOR gates. For the even 1's parity_out = 0 and for odd 1's parity_out = 1. The parity checker is shown in Fig. 5.18 The parity checker.

If each XOR gate has 0.1nsec propagation delay then the design has the 0.1 * 3 = 0.3nsec propagation delay.

We can optimize the delay by rearranging the logic to form the two stages as shown below (Fig. 5.19 The grouping of the XOR terms) and it is called as grouping of the terms to infer the parallel logic. This is useful to reduce the power and to

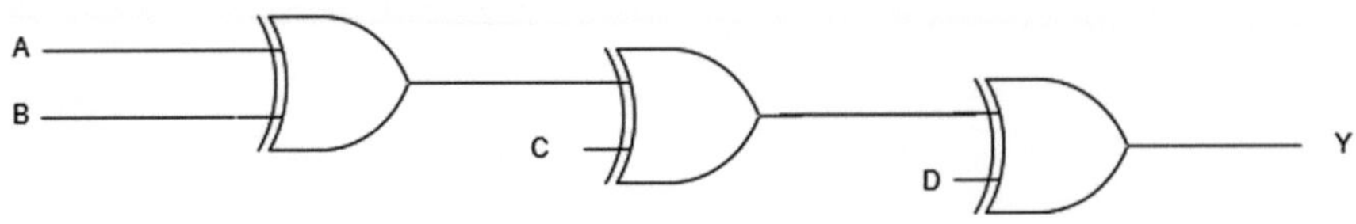

Fig. 5.18 The parity checker

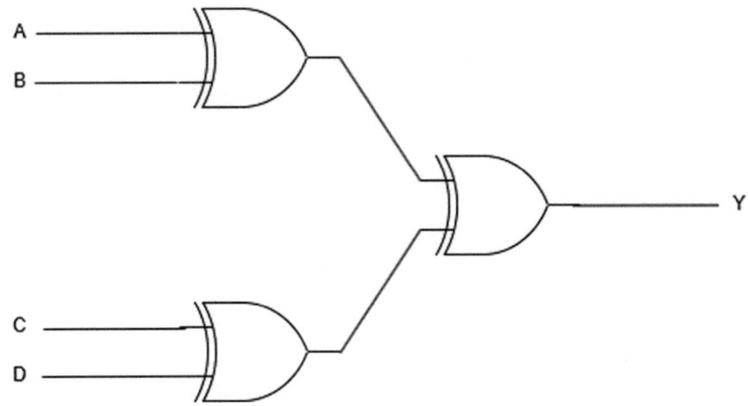

Fig. 5.19 The grouping of the XOR terms

improve the speed of the design. The overall propagation delay is 0.1nsec * 2 = 0.2nsec.

5.9 Practical Scenarios

In the subsequent chapters we will discuss the synchronous and asynchronous sequential circuits. Consider the scenario to design the sequence of one-cold encoding then we can use the following design.

1. The 2-bit binary synchronous counter
2. The 2:4 decoder having active low output (Table 5.10).

Table 5.10 One-cold encoding

CLK	Q1	Q0	Y3	Y2	Y1	Y0
1	0	0	1	1	1	0
2	0	1	1	1	0	1
3	1	0	1	0	1	1
4	1	1	0	1	1	1

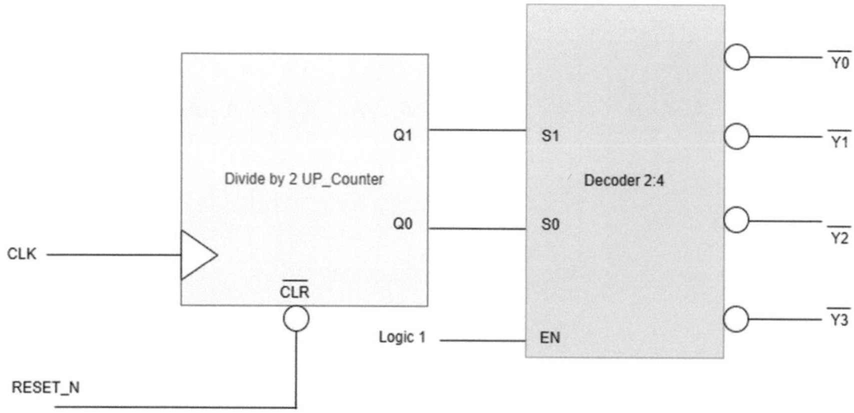

Fig. 5.20 The logic diagram of the sequence generator (one-cold encoding)

As shown in Fig. 5.20 The logic diagram of the, the design generates the one-cold encoding output 1110, 1101,1011,0111. Here one of the output bits is logic zero and it is one-cold encoding.

The decoders are used to select one of the memory or IO devices depending on the address. Consider the memory map shown below (Table 5.11: The memory mapping table).

Consider the ROM and RAM of 8Kbyte, ROM starting address is 0000H and end address is 1FFFH, The RAM #0 starting address is 2000H and end address 3FFFH, RAM#1 starting address is 4000H and end address is 5FFF hex.

To select one of the memory locations of the 8Kbyte * 16bit memory we need to have the 13 address lines (A12 to A0) and 16 data lines(D15-D0). To select the memory device one at a time we can use the A15, A14, A13 lines (Table 5.12).

As shown in the Table, A15 = 0 and used as enable input and A14 and A13 are connected to the S1, S0 lines of the decoder respectively. For the A14,A13 = 00 the ROM is selected, for the A14A13 = 01 RAM #0 is selected and for the A14A13 = 10 RAM #1 is selected.

The design using the 2:4 decoder select circuit is shown in the Fig. 5.21.

Let us use the techniques discussed in the Chapter 1 to Chapter 5 in subsequent chapters to design the area optimized sequential circuits and processor data path logic.

5.10 Important Highlights

Following are few of the important points to summarize this chapter.

1. The cascade chain of multiplexers incurs more delay in the design.

Table 5.11 The memory mapping table

Device	A15	A14	A13	A12	A11	A10	A9	A8	A7	A6	A5	A4	A3	A2	A1	A0
0000H	0	0	0	0	0	0	0	0	0	0	0	0	0	0	0	0
1FFFH	0	0	0	1	1	1	1	1	1	1	1	1	1	1	1	1
2000H	0	0	1	0	0	0	0	0	0	0	0	0	0	0	0	0
3FFFH	0	0	1	1	1	1	1	1	1	1	1	1	1	1	1	1
4000H	0	1	0	0	0	0	0	0	0	0	0	0	0	0	0	0
5FFFH	0	1	0	1	1	1	1	1	1	1	1	1	1	1	1	1

2. The priority mux logic has more area and more propagation delay as compared to the parallel mux.
3. The input swapping depending on the late arrival or fast arrival in the data path can be used to balance the delays.

Table 5.12 The memory chip-select table

Memory Device	A15	A14	A13
ROM	0	0	0
RAM #0	0	0	1
RAM #1	0	1	0

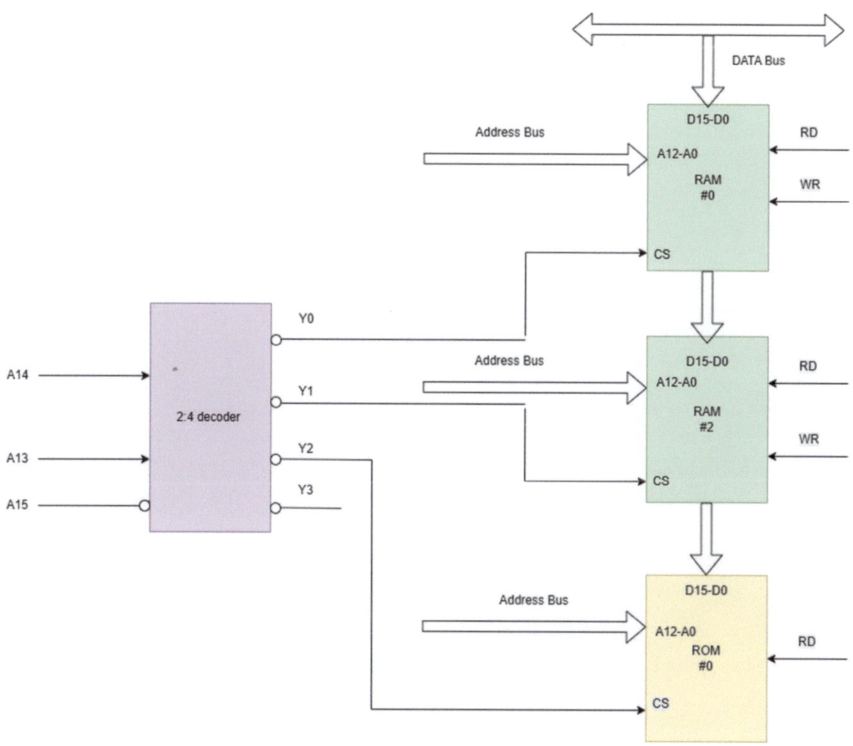

Fig. 5.21 The chip-select logic and memory organization

4. The decoders are used to select one of the memory or the peripheral device at a time depending on the status of the address inputs.
5. The larger decoders can be designed using the lower level hierarchy decoders.
6. The logic duplication techniques should be used to optimize the resources in the FPGA designs
7. The hazards are of two types, static and dynamic. Practically we should avoid the hazards in the design.
8. The grouping is used to improve the power and to reduce the propagation delays.

5.11 Puzzle

Design the positive level sensitive latch using the 2:1 mux. Design the negative level sensitive latch using 2:1 mux.

Chapter 6
Sequential Design

The design which uses the present input and past output is sequential design. The output of the sequential circuit is function of the present input and past output. The sequential design using flip-flops is discussed in this chapter and basic discussed are useful during the speed optimization discussion in the subsequent chapters. This chapter mainly focuses on the basics of the sequential design and how to design the optimized sequential circuits.

6.1 Level Sensitive Latches

The latches are the level sensitive and flip-flops are edge triggered elements and used to design the sequential circuits. For the sequential circuit the output is function of the present input and past output.

The latch output changes on the active level and flip-flop output on the active edge.

The positive and negative level sensitive latches, positive and negative edge triggered flip-flops are discussed in this section.

Throughout this manuscript we will use the D latches and D flip-flops. As we have single input D the design has minimum area and for the DFT friendly designs we use the D flip-flops and D latches with the time borrowing concept.

6.1.1 Positive Level Sensitive Latch

The latches are transparent during the active level. For the positive level sensitive latch, the active level is logic high that is positive level. During the positive level the latch output is same as the latch input.

V. Taraate, *Advanced Digital Design Techniques*,
https://doi.org/10.1007/978-981-95-4510-0_6

Table 6.1 Truth table of the positive level sensitive D latch

LE	Data input (D_IN)	Output (Q_OUT)
1	**0**	**0**
1	**1**	**1**
0	**X**	**Hold the previous output**

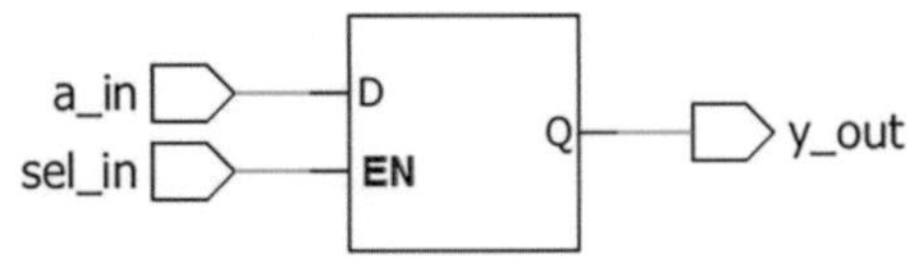

Fig. 6.1 Positive level sensitive D latch

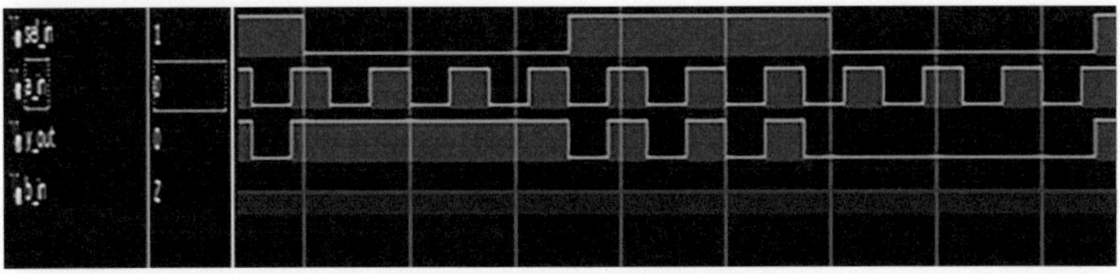

Fig. 6.2 Timing diagram of the positive level sensitive latch

During the inactive level that is logic 0 or negative level the latch holds the previous output.

The truth table of the positive level sensitive latch is shown in Table 6.1.

The D latch which is positive level sensitive is shown in Fig. 6.1.

The timing diagram of the positive level sensitive D latch is shown in Fig. 6.2. Timing diagram of the positive level sensitive latch. The latch output Q_OUT = D_In for the LE = 1 otherwise it holds the previous output.

6.1.2 Negative Level Sensitive D Latch

The latches are transparent during the active level. For the negative level sensitive latch, the active level is logic zero that is negative level. During the negative level the latch output is same as the latch input.

During the inactive level that is logic 1 or positive level the latch holds the previous output.

The truth table of the negative level sensitive latch is shown in Table 6.2.

The D latch which is negative level sensitive is shown in Fig. 6.3.

Table 6.2 The truth table of the negative level sensitive D latch

LE	Data input (D_IN)	Output (Q_OUT)
0	**0**	**0**
0	**1**	**1**
1	**X**	**Hold the previous output**

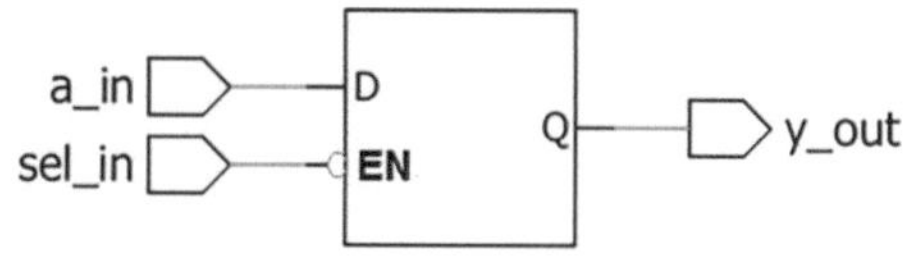

Fig. 6.3 Negative level sensitive D latch

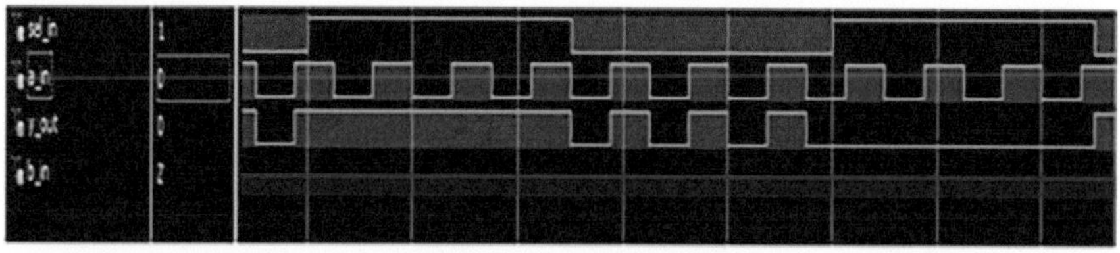

Fig. 6.4 Timing diagram of the negative level sensitive latch

The timing diagram of the negative level sensitive D latch is shown in Fig. 6.4. Timing diagram of the negative level sensitive latch. The latch output Q_OUT = D_In for the LE = 0 otherwise it holds the previous output.

6.2 D Flip-Flops: Edge Sensitive Elements

As discussed, we have the positive edge triggered flip-flops and negative edge triggered flip-flops. The flip-flops are edge sensitive and output of flip-flop is stable for one clock cycle as the input is captured on the edges.

6.2.1 Positive Edge Sensitive D Flip-Flop

The positive edge sensitive D flip-flop is designed by using the two latches in the cascade.

The output of the negative level sensitive latch is given to the positive level sensitive latch. Both latches uses the CLK as latch enable.

The positive edge sensitive D flip-flop is shown in Fig. 6.5.

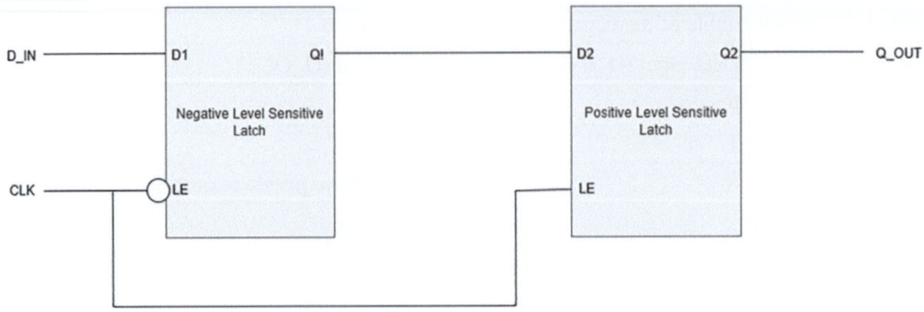

Fig. 6.5 Positive edge triggered D flip-flop

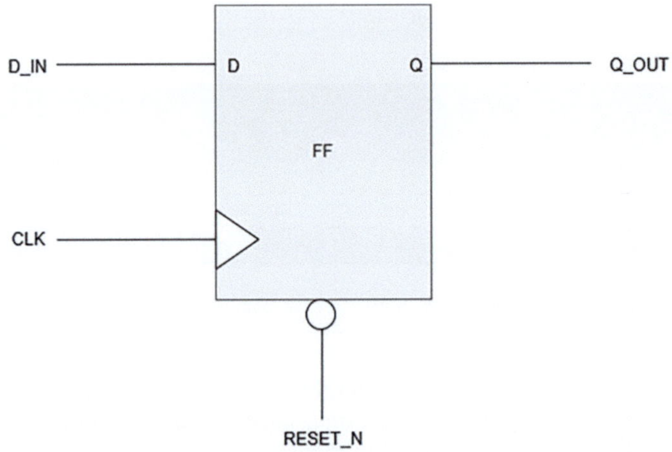

Fig. 6.6 Positive edge sensitive D flip-flop symbol

Throughout this manuscript we will use the following symbol to represent the positive edge sensitive D flip-flop (Fig. 6.6).

The timing diagram of the positive edge triggered D flip-flop is shown below and on the rising edge of the clock the flip-flop output Q_OUT = D_IN otherwise it will hold the previous output value (Fig. 6.7).

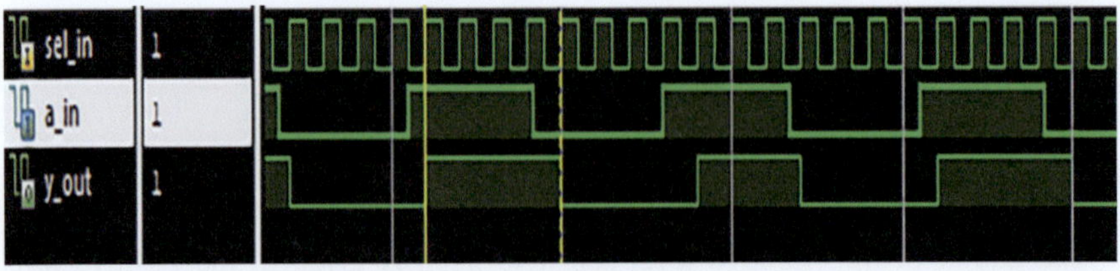

Fig. 6.7 Timing diagram of the positive edge sensitive D flip-flop

6.2.2 Negative Edge Sensitive D Flip-Flop

The output of the positive level sensitive latch is given as input to the negative level sensitive latch.

The negative edge sensitive D flip-flop is shown in Fig. 6.8.

We will use the following schematic of the negative edge sensitive throughout the manuscript (Fig. 6.9).

The timing diagram of the negative edge sensitive flip-flop is shown in Fig. 6.10 Timing diagram of the negative edge sensitive D flip-flop. As shown the output of flip-flop is stable for one clock cycle and is captured on the negative level.

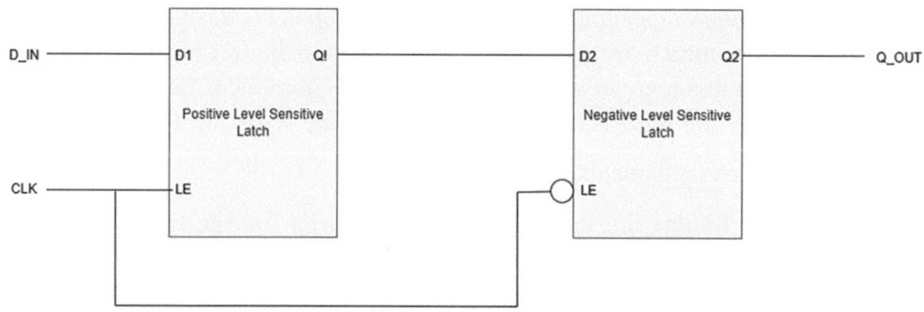

Fig. 6.8 Negative edge triggered D flip-flop

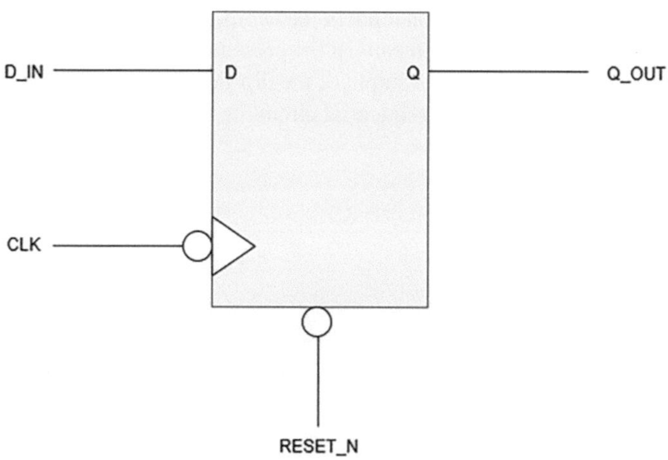

Fig. 6.9 Negative edge sensitive D flip-flop symbol

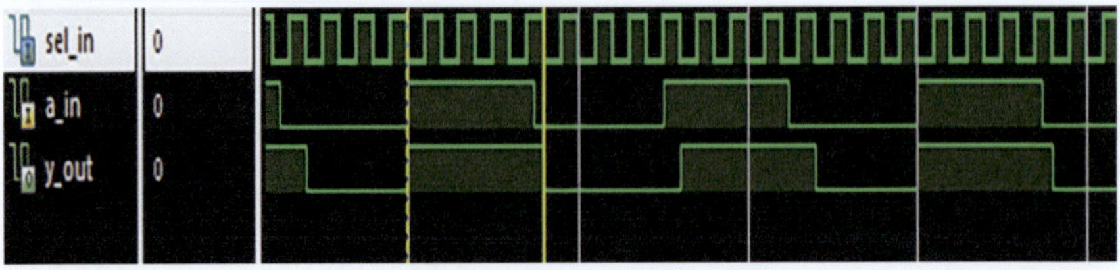

Fig. 6.10 Timing diagram of the negative edge sensitive D flip-flop

6.3 Frequency Divider

As D flip-flop has the single input, the D flip-flops are popular to design the sequential logic. It is easy to control the single D input as compared to the JK or SR flip-flops two inputs. Due to the less logic in the data path the area requirement for the sequential design using the D flip-flops is lesser as compared to SR or JK flip-flop.

What we need to understand about the D flip-flop?

For the sequential design using the D flip-flop we need to have the good understanding about the excitation inputs and excitation table. So let us try to discuss!

Consider the present state of the D flip-flop is Q and the next state of the D flip-flop is Q^+ (Table 6.3). The excitation table gives information about what should be the data input of the D flip-flop to get the next state output on the active edge of the clock. The excitation table of the D flip-flop is documented in Table 6.4.

As shown in Table 6.4 Excitation input of the D flip-flop the D input of the flip-flop is equal to the next state $\mathbf{Q^+}$. That means, if the present state $Q = 0$ and the next state $\mathbf{Q^+} = 1$ then to get the next state as output of the flip-flop the data input of flip-flop $D = 1$. That is $D = \mathbf{Q^+}$. During the sequential circuit designs and the FSM designs we

Table 6.3 State table of D flip-flop

Present state (Q)	Next state (Q^+)
0	0
0	1
1	0
1	1

Table 6.4 Excitation input of the D flip-flop

Present state (Q)	Next state (Q^+)	Excitation input (D)
0	0	0
0	1	1
1	0	0
1	1	1

Table 6.5 Excitation table of the D flip-flop

Present State (Q)	Next State (Q^+)	Excitation Input (D)
0	1	1
1	0	0

use the excitation table to deduce the combinational logic required at the data input of D flip-flop.

Now let us consider the design scenario

Consider the design requirement is to get the toggle output using the single D flip-flop. What we need to do is that let us document the entries in the excitation table and let us deduce the logic.

As shown in Table 6.5 Excitation table of the D flip-flop, the excitation input D $= Q^+$ which is the complement of the present state Q. So, $D = \overline{Q}$. So, the toggle flip-flop is designed by using the single D flip-flop and NOT gate. If the flip-flop uses the asynchronous active low reset, then in the reset path the clear input of the flip-flop is controlled by NOT gate.

The design is shown in Fig. 6.11.

6.4 Counters

Counters are used in the design to generate the predefined output sequence or as a divider network. The internally clock dividers are used to divide the input clock and to generate the desired output frequency.

We have two types of the counters

1. Synchronous counters
2. Asynchronous counters.

This chapter is useful to understand the design of the synchronous counters using minimum number of the design elements.

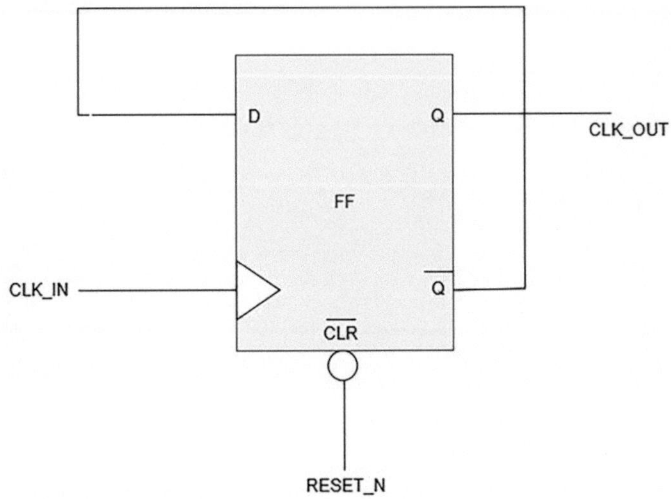

Fig. 6.11 Toggle D flip-flop

6.4.1 Synchronous Binary up Counter

Let us design the divide by 4 synchronous binary up counters. The divide by 4 is also called as MOD 4 because from the MSB of the counter the output frequency is input clock divide by 4.

Following are important steps used to design the counter

1. **Find the number of states to get the counter**

Number of states = 4.

2. **Find number of flip-flops**

Number of flip-flops $= n = \log_2 4 = 2$ We will use the positive edge sensitive D flip-flops.

3. **Reset strategy**

Let us use active low asynchronous reset input reset_*n*. For reset_$n = 0$ the counter holds the previous output. For the reset_$n = 1$ the counter output increments on the rising edge of the clock.

4. **Let us document the entries in the state table**

The state table of the MOD-4 synchronous binary up-counter is shown in Table 6.6.

Table 6.6 The state table of the MOD-4 binary up-counter

Present state (q1 q0)	Next state ($q1^+$ $q0^+$)
00	01
01	10
10	11
11	00

5. **Let us document the entries in the excitation table**

The excitation table consists of the information about the present state, next state, and excitation input (Table 6.7).

6. **Let us deduce the logic at the input of the D1, D0**

Let us use the present state q1, q0 as the inputs to get the logic at the D1, D0. Let us use the 2-varibale K-map (Figs. 6.12 and 6.13).

$$D1 = \overline{q1} \cdot q0 + \overline{q0} \cdot q1$$

$$D1 = q1 \oplus q0$$

$$D0 = \overline{q0}$$

7. **Let us sketch the logic**

As discussed in the previous steps we need to have two flip-flops and XOR gate to implement the MOD-4 synchronous binary up-counter (Fig. 6.14).

Table 6.7 The excitation table of the MOD-4 counter

Present state (q1 q0)	Next state ($q1^+$ $q0^+$)	Excitation input (D1 D0)
00	01	01
01	10	10
10	11	11
11	00	00

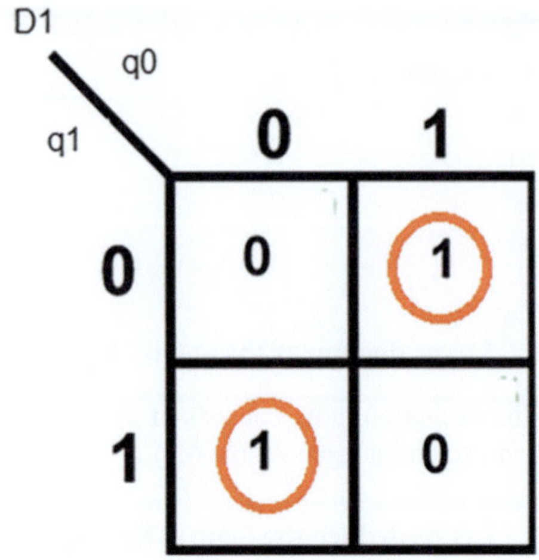

Fig. 6.12 The K-map to get the logic at D1

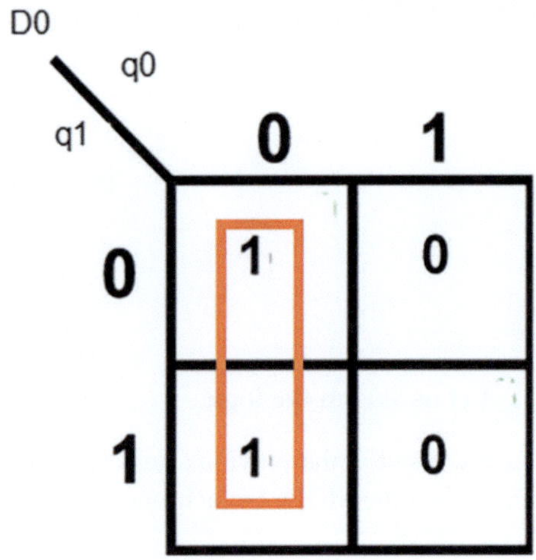

Fig. 6.13 The K-map to get the logic at D0

6.4.2 Synchronous Binary Down Counter

Now let us design the synchronous binary down counter that is to get the 2-bit output

1. **Find the number of states to get the counter**

Number of states = 4.

2. **Find number** of flip-flops

Number of flip-flops $= n = \log_2 4 = 2$ We will use the positive edge sensitive D flip-flops.

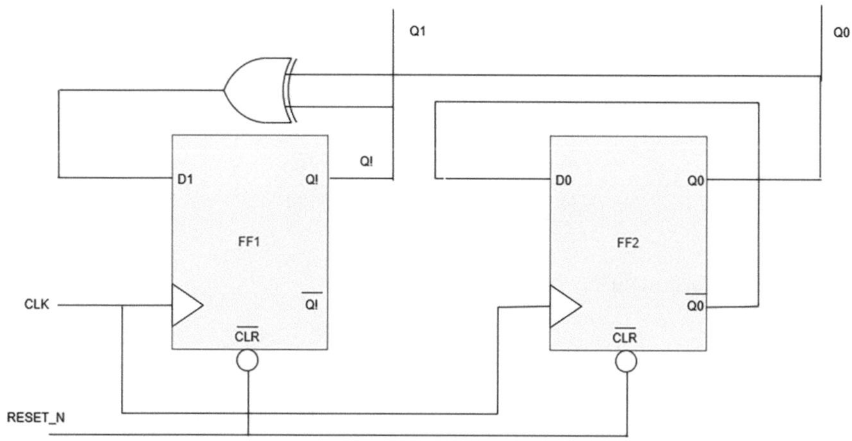

Fig. 6.14 2-bit synchronous binary up-counter

3. **Reset strategy**

Let us use active low asynchronous reset input reset_*n*. For reset_$n = 0$ the counter holds the previous output. For the reset_$n = 1$ the counter output decrements on the rising edge of the clock.

4. **Let us document the entries in the state table**

The state table of the MOD-4 synchronous binary down-counter is shown in Table 6.8.

5. **Let us document the entries in the excitation table**

The excitation table consists of the information about the present state, next state, and excitation input (Table 6.9).

6. **Let us deduce the logic at the input of the D1, D0**

Let us use the present state q1, q0 as the inputs to get the logic at the D1, D0. Let us use the 2-varibale K-map (Figs. 6.15 and 6.16).

$$D1 = \overline{q1} \cdot \overline{q0} + q1 \cdot q0$$

Table 6.8 The state table of the MOD-4 binary down-counter

Present state (q1 q0)	Next state ($q1^+$ $q0^+$)
11	10
10	01
01	00
00	11

Table 6.9 The excitation table of the MOD-4 down counter

Present state (q1 q0)	Next state ($q1^+$ $q0^+$)	Excitation input (D1 D0)
11	10	10
10	01	01
01	00	00
00	11	11

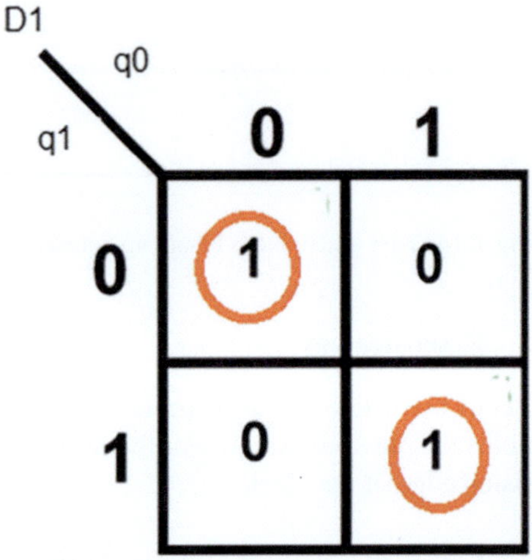

Fig. 6.15 The K-map to get the logic at D1

$$D1 = \overline{q1 \oplus q0}$$

$$D0 = \overline{q0}$$

7. **Let us sketch the logic**

As discussed in the previous steps we need to have two flip-flops and XNOR gate to implement the MOD-4 synchronous binary down-counter (Fig. 6.17).

6.4.3 Gray Counters

Now let us try to use the design foundation discussed in the previous sections to design the synchronous gray counter. As name itself indicates the counter has the 4 states. so, s1, s2,s3 and for the gray counter design using the following steps.

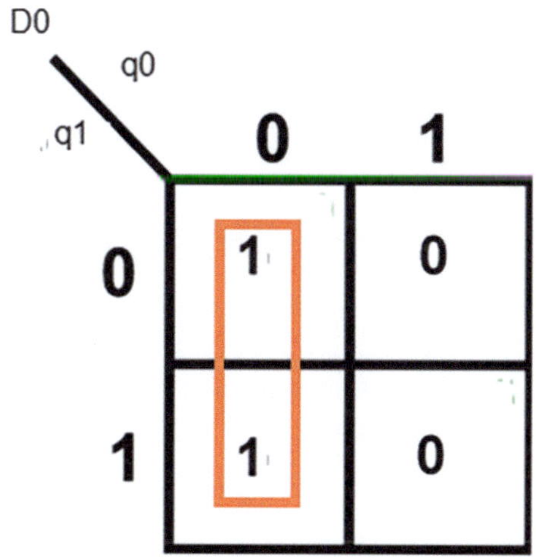

Fig. 6.16 The K-map to get the logic at D0

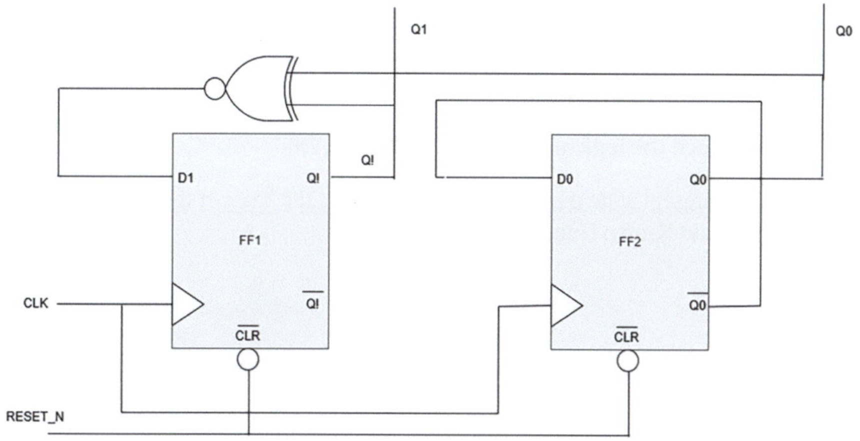

Fig. 6.17 2-bit synchronous binary down-counter

1. **Find the number of states to get the counter**

Number of states = 4.

2. **Find number of flip-flops**

Number of flip-flops = $n = \log_2 4 = 2$ We will use the positive edge sensitive D flip-flops.

Table 6.10 The state table of the 2-bit gray counter

Present state (q1 q0)	Next state ($q1^+$ $q0^+$)
00	01
01	11
11	10
10	00

3. **Reset strategy**

Let us use active low asynchronous reset input reset_n. For reset_$n = 0$ the counter holds the previous output. For the reset_$n = 1$ the counter output decrements on the rising edge of the clock.

4. **Let us document the entries in the state table**

The state table of the synchronous gray counter is shown in Table 6.10.

5. **Let us document the entries in the excitation table**

The excitation table consists of the information about the present state, next state, and excitation input (Table 6.11).

6. **Let us deduce the logic at the input of the D1, D0**

Let us use the present state q1, q0 as the inputs to get the logic at the D1, D0. Let us use the 2-varibale K-map (Figs. 6.18 and 6.19).

$$\mathrm{D1} = \mathrm{q0}$$

$$\mathrm{D0} = \overline{\mathrm{q1}}$$

7. **Let us sketch the logic**

As discussed in the previous steps we need to have two flip-flops to implement the synchronous gray counter (Fig. 6.20).

Table 6.11 The excitation table of the 2-bit gray counter

Present state (q1 q0)	Next state ($q1^+$ $q0^+$)	Excitation input (D1 D0)
00	01	01
01	11	11
11	10	10
10	00	00

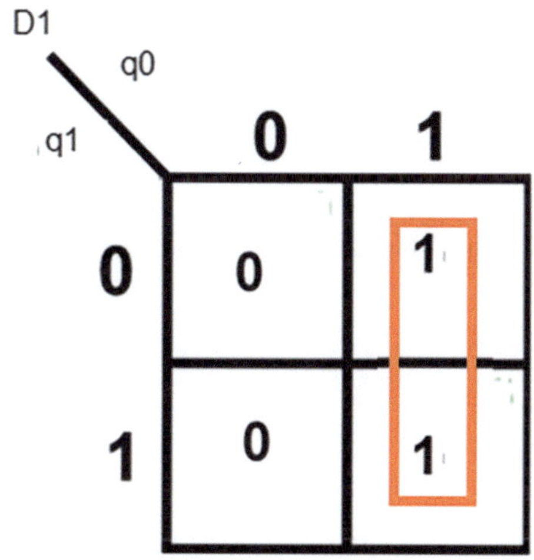

Fig. 6.18 The K-map to get the logic at D1

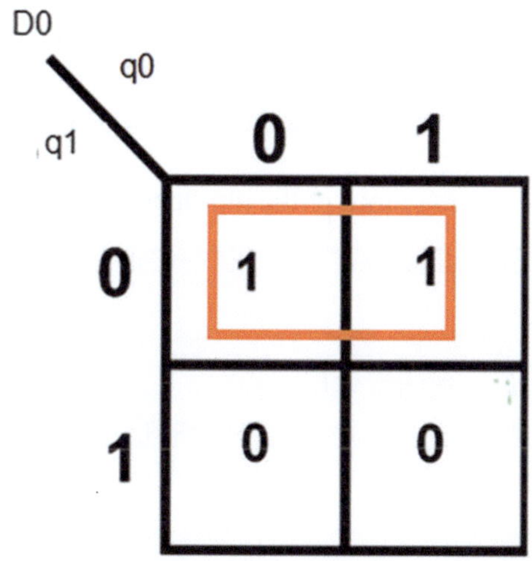

Fig. 6.19 The K-map to get the logic at D0

6.5 Ring Counter

As name indicates the ring counter generates an output sequence as 8,4,2,1,8,4…. The counter can be designed very quickly using the state table. So let us discuss the steps to design the ring counter.

1. **Find the maximum count**

For the counter to have output as 8, 4, 2, 1 the maximum count value is 8.

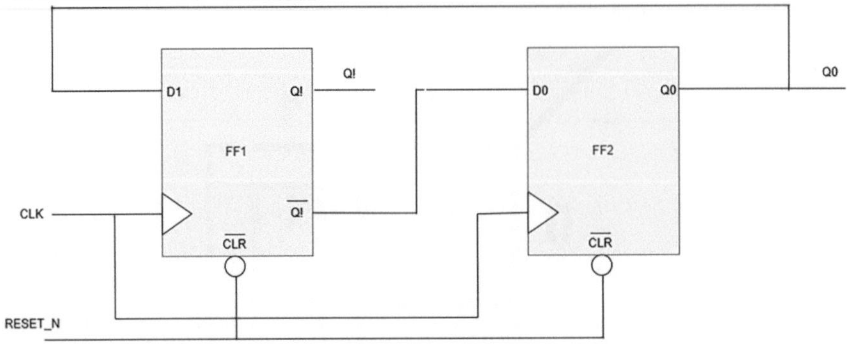

Fig. 6.20 2-bit gray counter

2. **Use the maximum count to find the number of flip-flops required**

In the binary 8 is represented as 1000. Hence the number of flip-flops required are equal to 4.

3. **Document the state table entries**

The counter has four states and the entries are documented in Table 6.12.

4. **Document the excitation input of the flip-flops**

Let us document the excitation input to get the next state count value (Table 6.13).

Table 6.12 The 4-bit ring counter state table

Present state (q3 q2 q1 q0)	Next state ($q3^+$ $q2^+$ $q1^+$ $q0^+$)
1000	0100
0100	0010
0010	0001
0001	1000

Table 6.13 The 4-bit ring counter excitation table

Present state (q3 q2 q1 q0)	Next state ($q3^+$ $q2^+$ $q1^+$ $q0^+$)	Excitation input (D3 D2 D1 D0)
1000	0100	0100
0100	0010	0010
0010	0001	0001
0001	1000	1000

5. **Observe the present state and next state to get the desired value as excitation to data input of flip-flops**

As shown the relationship between the present state and next state is, on the rising edge of the clock the output of the ring counter shifts by 1-bit. So, using this let us get the Boolean equations to have D3, D2, D1, D0 inputs (Table 6.14).

6. **Boolean equations**

Let us get the Boolean equations

$$D3 = Q0$$

$$D2 = Q3$$

$$D1 = Q2$$

$$D0 = Q1$$

7. **Let us sketch the logic**

The design of the 4-bit ring counter is shown in Fig. 6.21.

Table 6.14 Excitation table to deduce the expressions

Present State (q3 q2 q1 q0)	Next State ($q3^+$ $q2^+$ $q1^+$ $q0^+$)	Excitation Input (D3 D2 D1 D0)
1000	0100	0100
0100	0010	0010
0010	0001	0001
0001	1000	1000

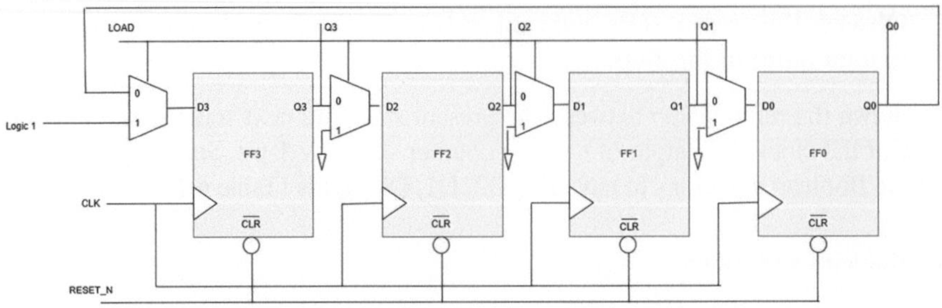

Fig. 6.21 The 4-bit ring counter design

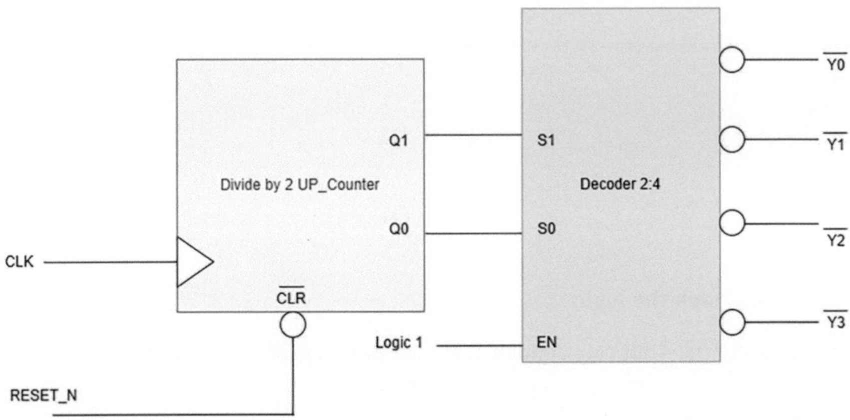

Fig. 6.22 The ring counter using the decoder and mod 4 counter

The design using the synchronous 2-bit binary up counter and the 2:4 decoder is shown in Fig. 6.22.

6.6 Twisted Ring Counter

As Johnson counter is twisted ring counter and used to generates an output sequence as 0, 8, 12, 14, 15, 7, 3, 1, 0…. The counter can be designed very quickly using the state table. So let us discuss the steps to design the ring counter.

1. **Find the maximum count**

For the counter to have output as 0, 8, 12, 14, 15, 7, 3, 1, 0… the maximum count value is 15.

Table 6.15 The 4-bit twisted ring counter state table

Present state (q3 q2 q1 q0)	Next state ($q3^+$ $q2^+$ $q1^+$ $q0^+$)
0000	1000
1000	1100
1100	1110
1110	1111
1111	0111
0111	0011
0011	0001
0001	0000

2. **Use the maximum count to find the number of flip-flops required**

In the binary 15 is represented as 1111. Hence the number of flip-flops required are equal to 4.

3. **Document the state table entries**

The counter has four states and the entries are documented in Table 6.15.

4. **Document the excitation** input of the flip-flops

Let us document the excitation input to get the next state count value (Table 6.16).

5. **Observe the present state and next state to get the desired value as excitation to data input of flip-flops**

As shown the relationship between the present state and next state is, on the rising edge of the clock the output of the ring counter shifts by 1-bit and the MSB of the counter is complement of the LSB output. So, using this let us get the Boolean equations to have D3, D2,D1, D0 inputs (Table 6.17).

Table 6.16 Excitation table of twisted ring counter

Present state (q3 q2 q1 q0)	Next state ($q3^+$ $q2^+$ $q1^+$ $q0^+$)	Next state (D3 D2 D1 D0)
0000	1000	1000
1000	1100	1100
1100	1110	1110
1110	1111	1111
1111	0111	0111
0111	0011	0011
0011	0001	0001
0001	0000	0000

Table 6.17 Excitation table to deduce the expressions

Present State (q3 q2 q1 q0)	Next State ($q3^+$ $q2^+$ $q1^+$ $q0^+$)	Next State (D3 D2 D1 D0)
0000	1000	1000
1000	1100	1100
1100	1110	1110
1110	1111	1111
1111	0111	0111
0111	0011	0011
0011	0001	0001
0001	0000	0000

6. **Boolean equations**

Let us get the Boolean equations

$$D3 = \overline{Q0}$$

$$D2 = Q3$$

$$D1 = Q2$$

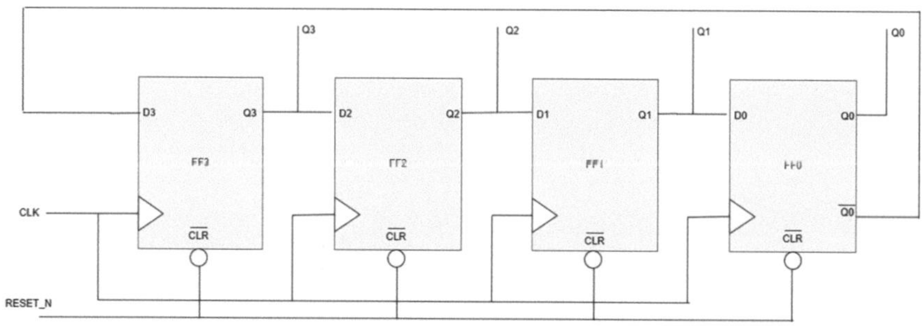

Fig. 6.23 The 4-bit Johnson counter design

$$D0 = \overline{Q1}$$

7. **Let us sketch the logic**

The design uses the four flip-flops and the counter is called as twisted ring counter (Fig. 6.23).

6.7 Important Highlights

Following are few of the important points to summarize this chapter.

1. The sequential design elements are latches and flip-flops.
2. In the sequential design an output is function of the present inputs and past outputs.
3. The D latch is level sensitive
4. The D flip-flop is edge sensitive
5. The latches are transparent during active level of the enable input.
6. The flip-flop samples data input on the active edge of the clock.
7. To design the counters and shift-registers we will use the flip-flops as they are edge sensitive.

Chapter 7
Timing Parameters and Optimization Scenarios

In the previous few chapters, we have discussed the combinational and sequential design elements. This chapter is useful to understand the timing parameters, insertion delays, timing paths and the slack calculations. The chapter is also useful to understand the setup and hold slack calculations and various design scenarios and how to fix the issues. The chapter is useful to understand the basics of multiple clock domain designs, glitch free clock mux design and the skews in the design.

7.1 D Flip-Flop Timing Parameters

The D flip-flop is edge sensitive and the rising edge sensitive D flip-flop is shown in Fig. 7.1.

The flip-flop timing parameters are setup time, hold time and clock to q delay.

1. **Setup Time**: It is denoted by tsu and it is defined as minimum amount of the time for which the data input should remain stable before arrival of the active clock edge.
2. **Hold Time**: It is denoted by th and is defined as the minimum amount of time for which the data input should maintain stable after active edge of the clock.
3. **Clock to Q delay**: It is denoted by tctoq or tpff and it is the propagation delay of the flip-flop that is time required for the flip-flop to generate valid output after arrival of the active edge of the clock.

As shown in the timing diagram the clock time period is TCLK and the tsu is setup time which is minimum time for which data is stable at launch as well at capture edge. The th = hold time, is the minimum amount of time for which data should be stable after arrival of the active edge of the clock.

That is during the window of the setup + hold the data should be stable. If the data changes during this window, then the timing violation in the design.

V. Taraate, *Advanced Digital Design Techniques*,
https://doi.org/10.1007/978-981-95-4510-0_7

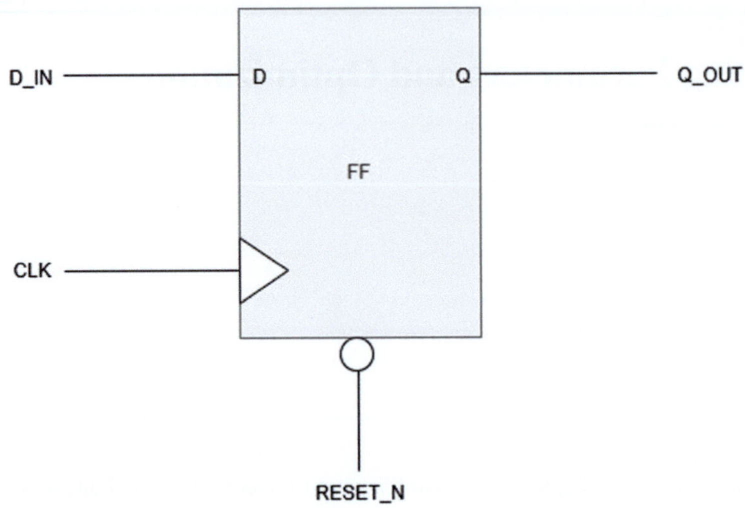

Fig. 7.1 D Flip-flop

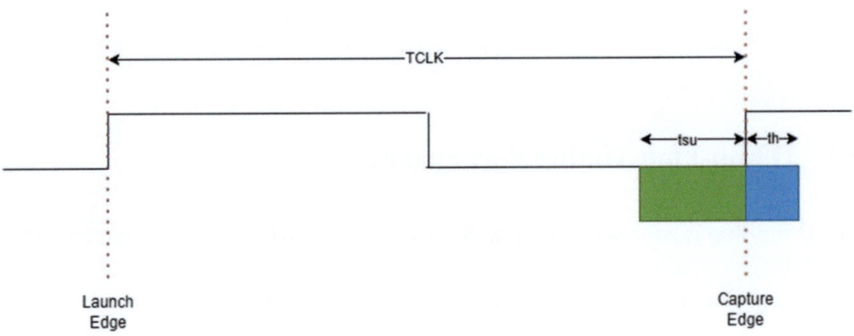

Fig. 7.2 The timing diagram of setup and hold window

7.2 Input Insertion Delays

In the design we have the insertion delays. As shown in Fig. 7.3 the combo logic in the data path inserts the combinational delay. The timing path is input to reg path and start point is DATA_IN, end point is D of the sequential element.

To avoid any timing violation the data at the D input of the flip-flop should be stable during the setup and hold window.

The combo delay is tcombo then the arrival time is AT = tcombo and required time is RT = Tclk-tsu. The RT should be greater than or equal to AT. This indicates no setup violations in the design.

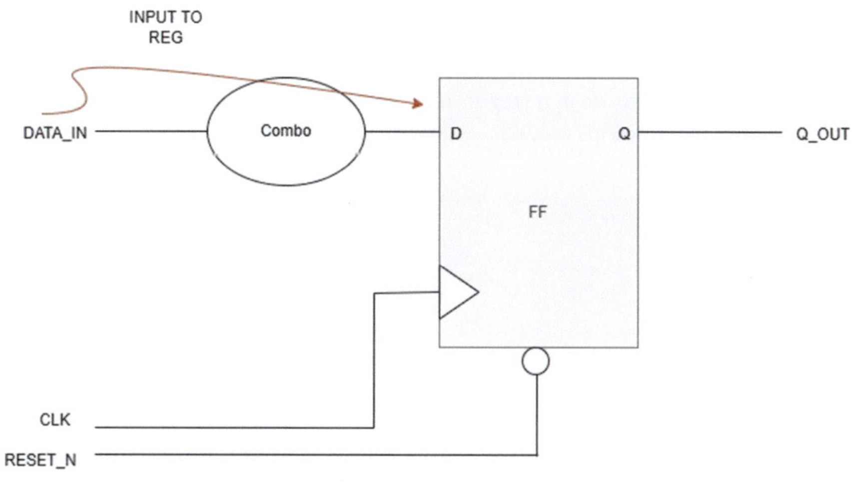

Fig. 7.3 Input insertion delay

7.3 Output Insertion Delays

At the output of the flip-flop the combo logic is placed and inserts the delays in the design. The path is called as the reg to output path.

As shown in Fig. 7.4. the data at the output of the flip-flop is available at tctoq and if we consider the tcombo delay the data at the output Q_OUT is available at tctoq + tcombo.

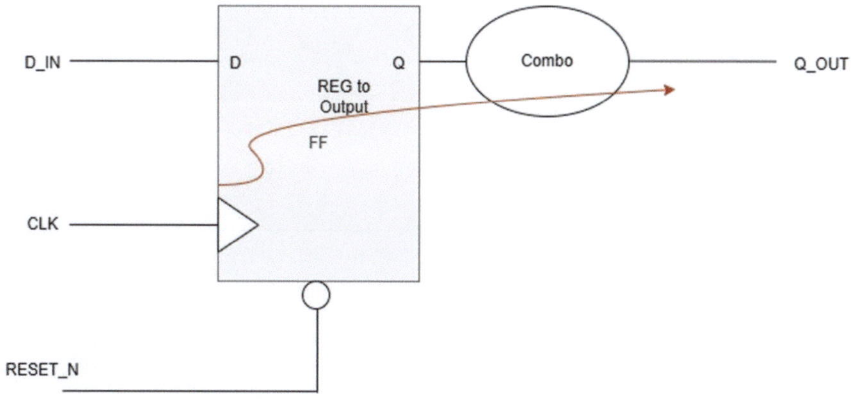

Fig. 7.4 Output insertion delays

7.4 Reg to Reg Path

Figure 7.5 Synchronous design (reg to reg path) shows the synchronous sequential circuit as both the flip-flops clock is derived from the common clock source.

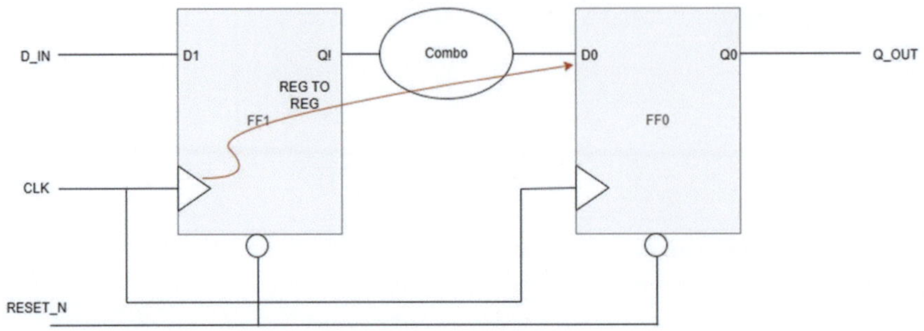

Fig. 7.5 Synchronous design (reg to reg path)

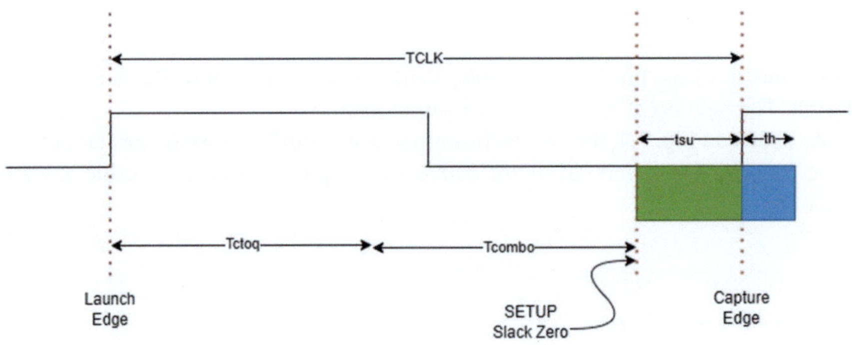

Fig. 7.6 The zero-setup slack

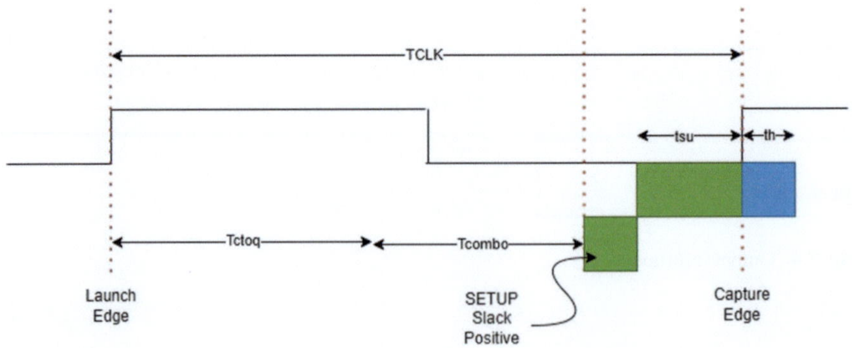

Fig. 7.7 The positive setup slack

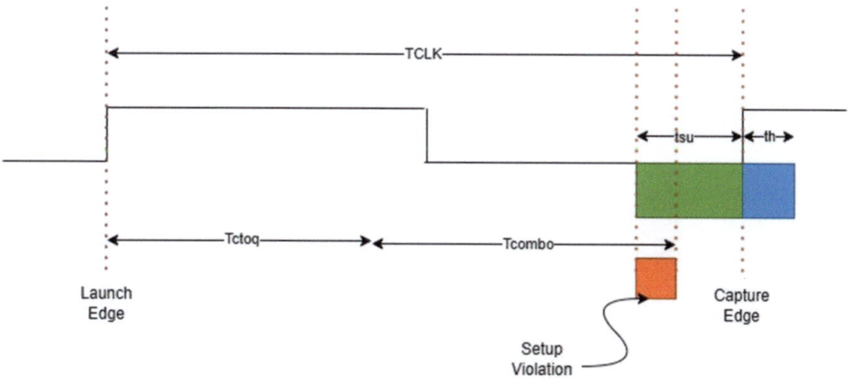

Fig. 7.8 The setup slack violation

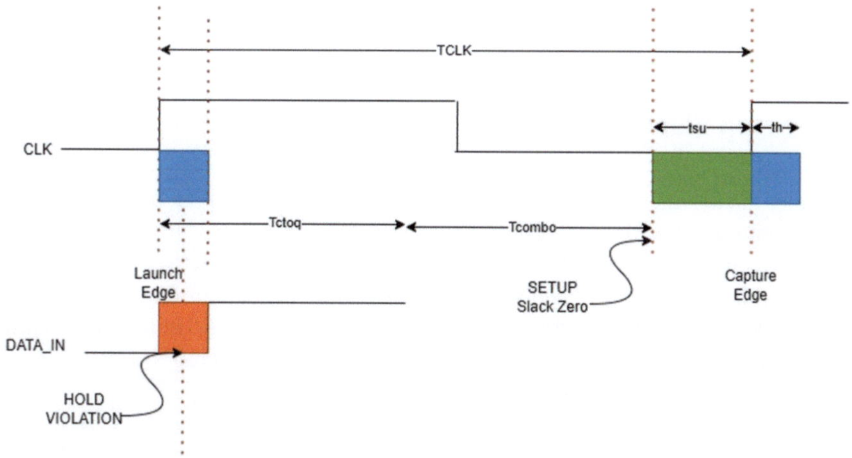

Fig. 7.9 The hold violations in the design

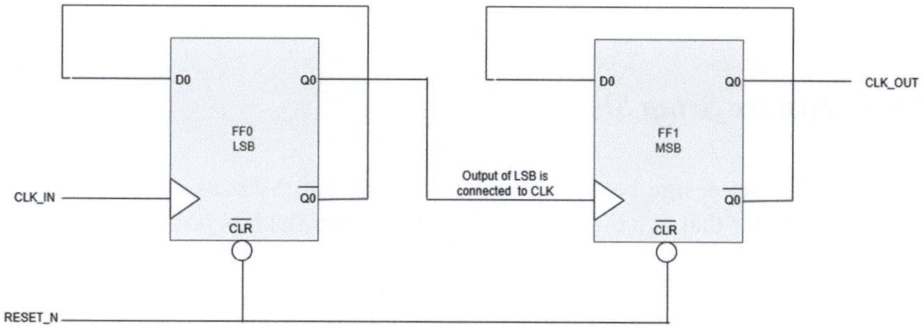

Fig. 7.10 The asynchronous counter

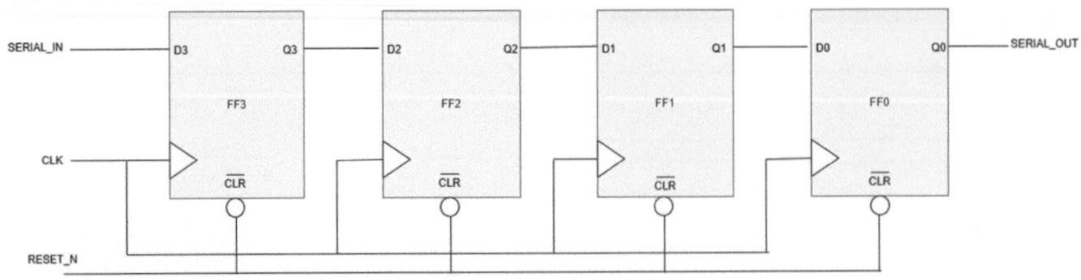

Fig. 7.11 The serial in serial out register

The timing path is called as reg to reg path and the RT which is required time should be greater than or equal to the arrival time AT to avoid the timing violations.

RT = Tclk-tsu.

AT = tctoq1 + tcombo. Here tcombo = 0

RT > = AT so no timing violations in the design.

7.5 Setup Slack

Consider the design shown in Fig. 7.5, the timing path is reg to reg path.

The setup slack is the difference between the RT and AT.

For the design shown below the RT = TCLK-tsu and AT = tctoq1 + tcombo.

So, setup slack is RT-AT and should be positive to meet the timing.

Consider the design has max operating frequency of 500 MHz and has timing parameters tsu = 0.5 ns, th = 0.25 ns, tcombo = 0.50 ns and tctoq = 1 ns.

So, RT = TCLK − 0.5 ns, AT = 1 ns + 0.5 ns = 1.5 ns.

Setup slack = RT − AT = TCLK − 0.5ns − 1.5 ns > = 0.

Here TCLK is reciprocal of the 500 MHz and it is 2 ns so setup slack is 2.0 ns–0.5 ns–1.5 ns = 0 and shown in Fig. 7.6. Hence design works at 500 MHz and no timing violation in design.

7.5.1 Positive Setup Slack

Now consider the design has max frequency of the 500 MHz as discussed, and if data arrives faster that is tcombo = 0.25 ns. The setup slack is positive as the data arriving is faster.

So, RT = TCLK − 0.5 ns, AT = 1 ns + 0.25 ns = 1.25 ns.

Setup slack = RT–AT = TCLK − 0.5 ns − 1.25 ns > = 0.

Here TCLK is reciprocal of the 500 MHz and it is 2 ns so setup slack is 2.0 ns-0.5 ns-1.25 ns = 0.25 ns and shown in the Fig. 7.7. Design met timing and no violation in the design.

7.5.2 *Negative Setup Slack*

Now consider the design has max frequency of the 500 MHz as discussed, and if data arrives slower that is tcombo = 0.60 ns. The setup slack is negative as the data arriving is slower.

So, RT = TCLK − 0.5 ns, AT = 1 ns + 0.60 ns = 1.60 ns.

Setup slack = RT–AT = TCLK − 0.5 ns − 1.60 ns > = 0.

Here TCLK is reciprocal of the 500 MHz and it is 2 ns so setup slack is 2.0 ns − 0.5 ns − 1.60 ns = − 0.10 ns and shown in Fig. 7.8. Design has not met timing and setup violation in the design.

7.6 Hold Violations

If the data changes during the hold window, then the design will have hold violation. As shown in the following figure to avoid the hold violation the data should not change in the hold window.

To avoid hold violations the AT should be greater than the RT that is AT = tctoq + tcombo should be greater than the th.

If we have hold of 0.25 ns and setup of the tcombo = 0.5 ns, tctoq = 1 ns then the AT = 1.5 ns is greater than the th = 0.25 ns and hence no hold violations.

When hold violation we can have. Consider the above design we will try to have hold of 0.25 ns and if the tpd + tcombo = 0.20 then the design will have the hold violation. Here the tpd which is tctoq delay of the flip-flop plus the tcombo delay is lesser than the th.

7.7 Asynchronous Circuits

The asynchronous designs we will not recommend as the clocks are not derived from the common clock source. Consider the asynchronous design shown in the figure 7.10. The design uses the toggle flip-flops which are designed by using the D flip-flop. Each rising edge toggle flip-flop is divide by 2 counters.

The LSB flip-flop receives the clock from the PLL and the output of the LSB flip-flop is used as the input to the MSB flip-flop.

If the input clock is 500 MHz, then the Q0 frequency is divide by 2 that is 250 MHz, the Q1 which is CLK_OUT frequency is again divided by two. The CLK_OUT is 125 MHz.

7.8 Practical Scenarios

Consider the following shift register shown in the Fig. 7.11. It has the serial input and has the serial output. The four stages are used. Let us consider the timing parameters as tsu = 0.5 ns, th = 0.25 ns, tctoq = 1 ns.

The setup slack calculation.

AT = tctoq = 1 ns.

RT = TCLK-tsu = TCLK − 0.5 ns.

Setup Slack = RT–AT = TCLK − 0.5 ns − 1 ns = TCLK − 1.5 ns > = 0.

TCLK = 1.5 ns.

Fmax is 1/1.5 ns which is equal to the 666.66 MHz.

The hold is 0.25 ns and as AT is tctoq = 1 ns there is no any hold violation in the design.

7.9 Clock Skew

The time difference between the arrival of the clock is called as clock skew. As shown in Fig. 7.12 The clock skew, the design uses the PLL to generate a clock. The clock latency is due to the clock buffers

1. FF0 receives clock after the delay of the 2 * tbuf. If we consider the delay of buffer as 100 ps then the clock will arrive at t0 + 100*2 ps = t0 + 200 ps
2. The FF1 receives the clock after the delay of the 3*tbuf. The clock at the FF1 arrives at the t0 + 3 * 100 ps = t0 + 300 ps
3. The clock skew is: t0 + 300 ps –(t0 + 200 ps) = 100 ps which is difference between the arrival time of the two clock edges.

In the subsequent chapter we will discuss the positive and negative clock skew.

7.10 Can We Control the Clock Polarity?

In most of the design we may need the polarity control to control the polarity of the clock. We can pass the clock as it is to the various functional blocks.

Consider the design has the SDR and DDR memories and requirement is that, the DDR should receive the 180-degree phase shifted clock to perform the operation.

In such scenarios we can use the XOR gate as a polarity logic. One of the inputs of the XOR gate is CLK input and another input should be polarity control. The details are listed in the Table 7.1.

The design is shown in Fig. 7.13 The clock polarity controlled by XOR gate.

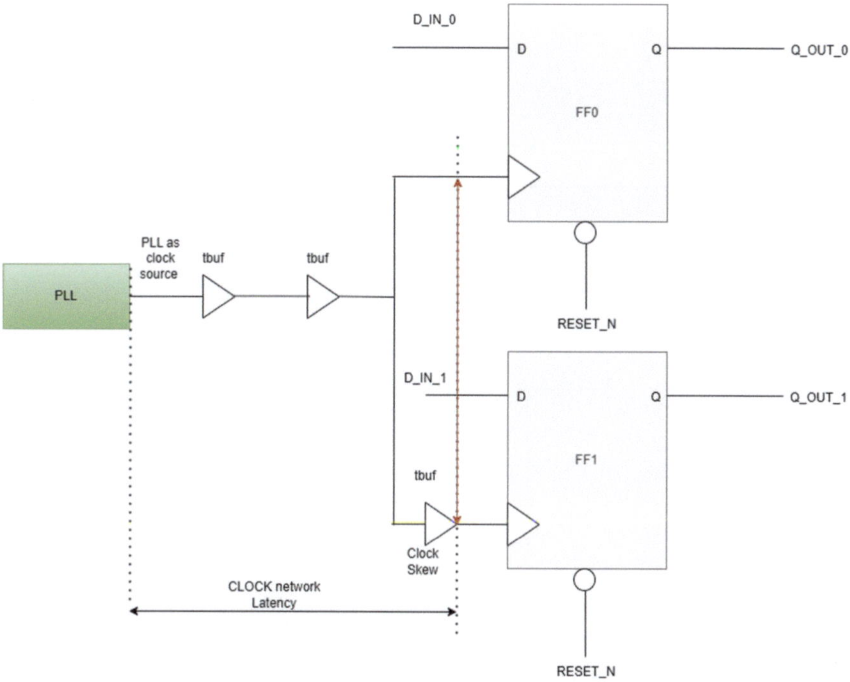

Fig. 7.12 The clock skew

Table 7.1 The polarity control logic

POL_CONTROL	Polarity logic output
0	CLK
1	The 180-degree phase shifted CLK

The timing diagram of the DDR clock is shown in Fig. 7.14 The timing diagram of the clock polarity control below. As shown for the POL_CONTROL = 0 the DDR CLK is same as CLK and for the POL_CONTROL = 1 the DDR CLK is complement of the CLK.

7.11 Clock Domains in the Design

Practically most of the designs are multiple clock domain designs and have more than one clock. Here the main issue is the metastability and data convergence. To resolve the issue, we need to deploy the synchronizers in the data and control path. The details of the synchronizers and synchronization techniques are discussed in Chapter 9.

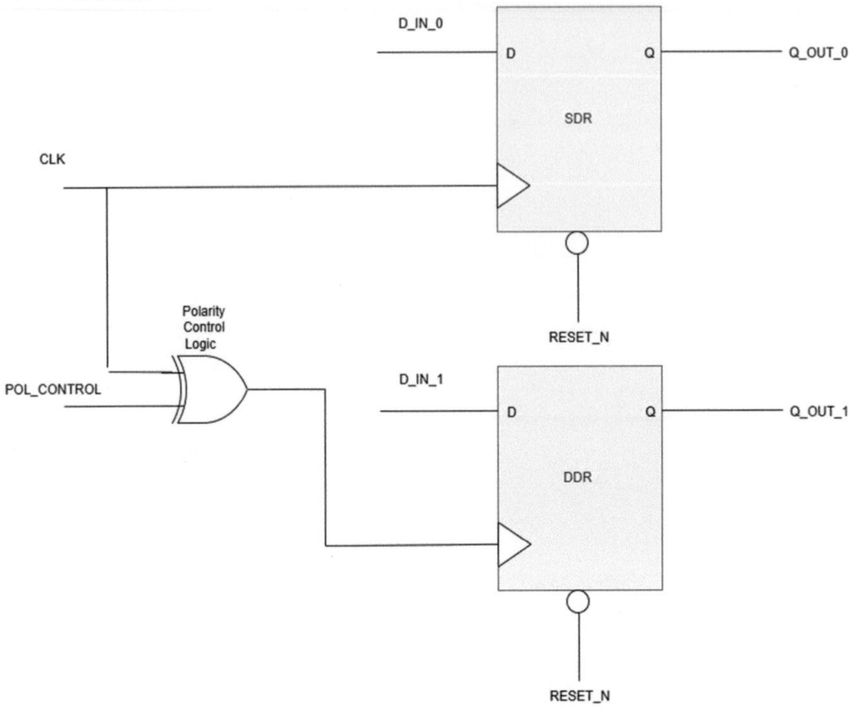

Fig. 7.13 The clock polarity controlled by XOR gate

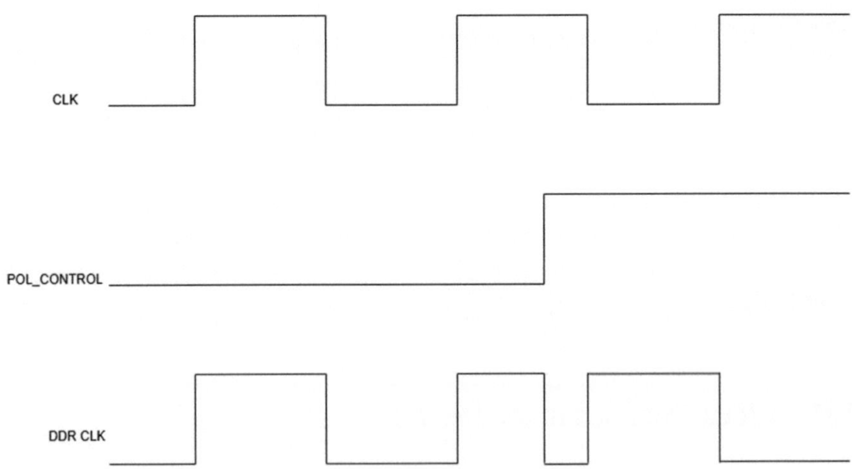

Fig. 7.14 The timing diagram of the clock polarity control

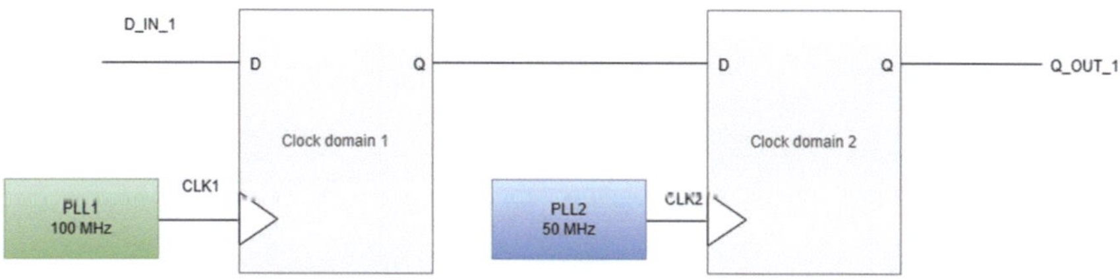

Fig. 7.15 The multiple clock domain designs

Metastability due to the clock domain crossing

Consider the design shown in the following Figure. The issues are at the clock domain crossing boundary.

The clock domain logic I operates at 100 MHz clock and clock domain logic II frequency is slower clock of 50 MHz.

As shown in Fig. 7.15 The multiple clock domain designs, the output from the clock domain I is connected as an input to the clock domain II. Here as both clocks are asynchronous in nature due to the phase difference between these clocks the output Q_OUT_1 goes into the metastable state. To avoid the metastability we need to deploy the synchronizers in the data and control path. It is not recommended to directly connect the output from one clock domain to another.

Control path synchronizers:

For the single bit signals to pass from the clock domain I to clock domain II we can deploy the level synchronizers in the control path.

The output from the clock domain I is sampled by the level synchronizer using the clock domain II clock and level synchronizer output is valid on the second clock cycle.

The output of the first flip-flop of the level synchronizer is always metastable and timing of this flip-flop should be disabled.

The level synchronizer deployed in the control path is shown in Fig. 7.16 The control path synchronizers.

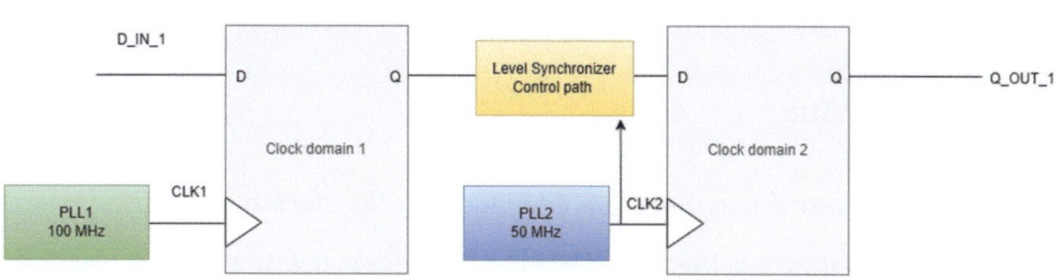

Fig. 7.16 The control path synchronizers

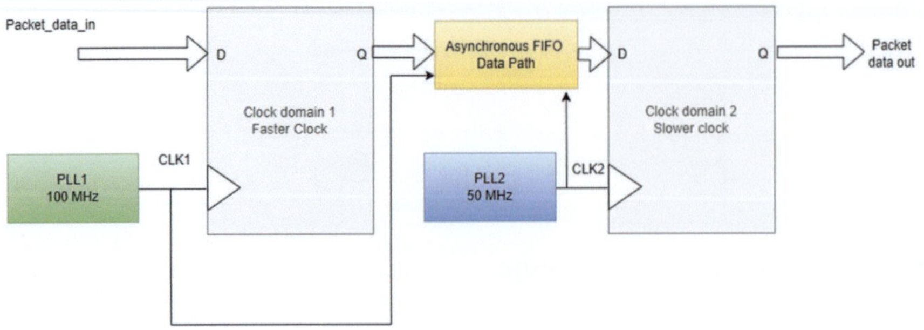

Fig. 7.17 The data path synchronizers

Data Path Synchronizer:

To pass the multi-bit data from faster clock domain to the slower clock domain we need to have the data path synchronizer. The FIFO Depth is dependent of the write clock and read clock.

As shown in Fig. 7.17 The data path synchronizers, the write clock here is 100 MHz, read clock is 50 MHz and if we need to pass 50 packets each of the 8-bits from faster clock domain to the slower clock domain then the FIFO depth is

$$\text{DEPTH} = \text{Burst} - \frac{\text{Burst} * \text{twrite}}{\text{tread}}$$

$$\text{Here twrite} = 1/100\ \text{MHz} = 10\ \text{ns}$$

$$\text{Tread} = 1/50\ \text{MHz} = 20\ \text{ns}$$

$$\text{Burst} = 50$$

$$\text{Therefore, the depth is}$$

$$\text{DEPTH} = 50 - \frac{50 * 10\,\text{ns}}{20\,\text{ns}}$$

$$\text{DEPTH} = 50 - 25 = 25$$

More details on the FIFO depth calculation are discussed in Chap. 9.

7.12 Clock Mux

Let us recapture our discussion from Chapter 2 on the clock mux.

The clock mux is shown in Fig. 7.18 The clock mux logic below and it has two clock inputs CLK_0, CLK_1 and to select one of the input the CLK_SELECT is used. The output of the clk mux is CLK_OUT.

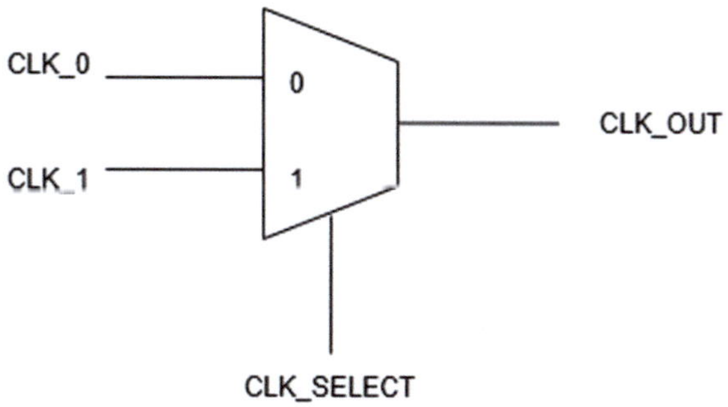

Fig. 7.18 The clock mux logic

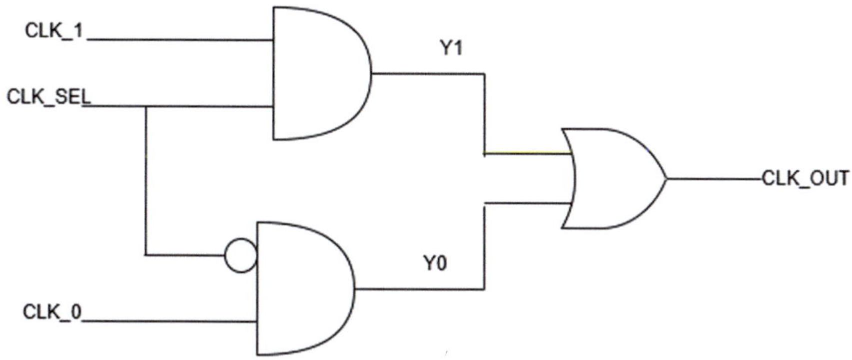

Fig. 7.19 The clock mux logic using gates

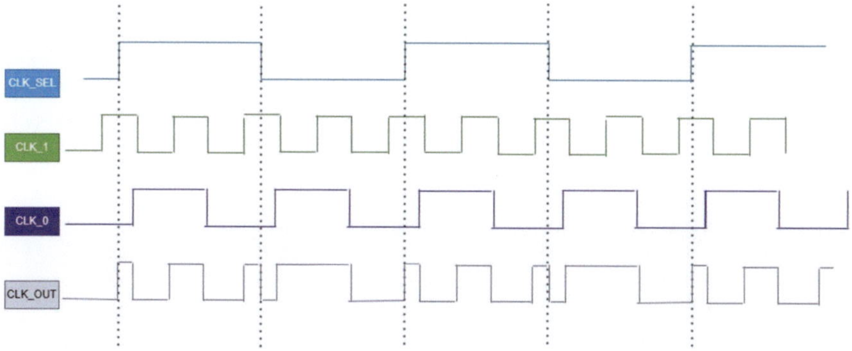

Fig. 7.20 The timing diagram of clock mux

As shown in the Table 7.2 The clock selection table below the design has CLK_OUT and for the CLK_SEL = 1 the CLK_OUT = CLK_1, for the CLK_SEL = 0 the CLK_OUT = CLK_0.

Table 7.2 The clock selection table

CLK_SEL	CLK_OUT
0	CLK_0
1	CLK_1

The logic of the clock mux is shown in the Fig. 7.19 and is prone to the glitches due to the asynchronous clock select input. And it is not recommended to use this type of clock muxing in the design.

The design is prone to the glitches and we need to have the glitch free clock mux. The timing diagram is shown in Fig. 7.20. Here the CLK_SEL is asynchronous input and CLK_1, CLK_0 are two inputs of the clk mux.

The clock mux design using the synchronizers and the OR gate is discussed subsequently to have the glitch free output.

Glitch free clock mux

The main objective of the clock mux cell is to select one of the clocks depending on the status of the clock select. The output of the clock mux should be glitch free and designer should take care to have the clean clock for the ASIC designs.

If we use the combinatorial logic as clock mux then the design is prone to glitches and we have discussed this in the above section. To avoid glitches let us use the flip-flops as shown in the figure and let us feedback output of the flip-flop to the input logic gates.

Here for the CLK_SEL = 0 the output of clock mux is CLK_0 and for the CLK_SEL = 1 the output of the clock mux is CLK_1.

The glitch free clock mux design is shown in Fig. 7.21 The glitch free clock mux logic.

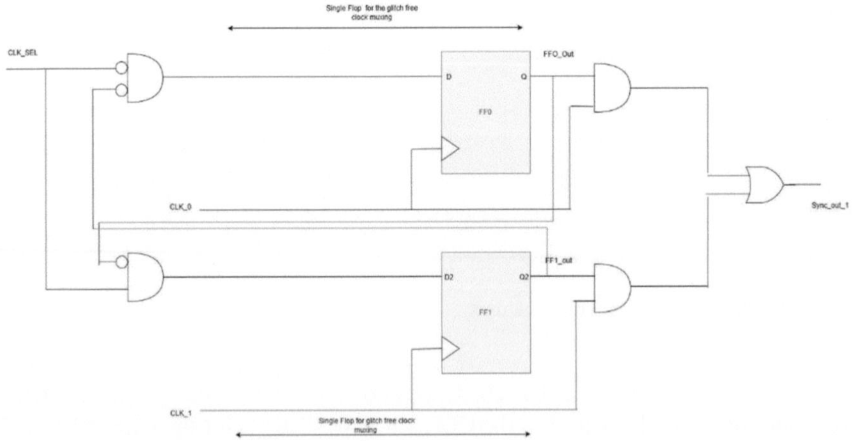

Fig. 7.21 The glitch free clock mux logic

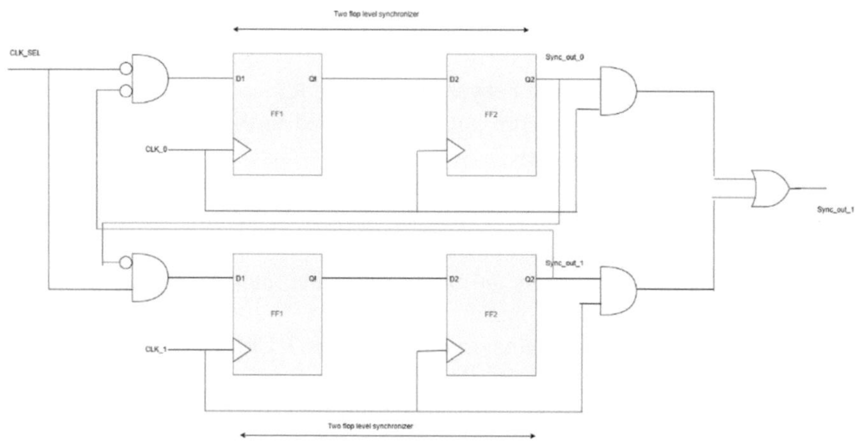

Fig. 7.22 The glitch free clock mux logic having level synchronizers

If we carefully observe still, there is issue. The issue is due to asynchronous CLK_SEL and if the CLK_SEL transits near to the edges of the CLK_0, CLK_1 then we will not be able to get the clean clock.

This issue can be resolved using the level synchronizer in the path of the AND gates.

7.13 Level Synchronizer to Avoid the Glitches

We can deploy the two-flop level synchronizer to avoid the glitches in the clock mux output. We will get the clean clock. As the CLK_SEL is asynchronous in nature and if it transits near to the clock edges then also the output of the level synchronizer is stable for the two clock cycles. This resolves the issue and output of the clock mux is clean clock.

The design is shown in Fig. 7.22 The glitch free clock mux logic having level synchronizers.

7.14 Important Highlights

Following are few of the important points to summarize this chapter.

1. The setup slack is difference between the RT and AT
2. The setup slack is positive then the design doesn't have any timing violations.
3. The setup slack is zero indicates no violations in the design, the setup slack is negative indicates the setup violations

4. For the reg to reg path the maximum frequency is calculated by using the $f\max = \frac{1}{tpff + tcombo + tsu}$
5. The hold slack is difference between the AT and RT
6. The asynchronous circuits are not recommended in the design as they have maximum propagation delay.
7. The skew in the design is difference in the arrival time of the clocks.
8. The control path synchronizers are used to pass the control signals from the faster clock domain to slower clock domain
9. The data path synchronizers are used to pass the multi-bit packets from the faster to the slower clock domain.
10. The clock mux using combinatorial logic that is 2:1 mux is prone to glitches and not recommended in the ASIC design
11. The glitch free clock mux we can design using the level synchronizers and logic gates.

Chapter 8
Timing Analysis and Speed Optimization

In the previous chapter we have discussed the timing parameters and how to find the maximum frequency for any design. This chapter is useful to understand the timing details, timing analysis and the speed optimization, improvement techniques. The chapter covers the practical scenarios for the register balancing, pipelining, and the speed improvement of the design.

8.1 Clock Skew

As discussed in the previous chapter the skew is the difference in the arrival time of two clocks. As shown in the Fig. 8.1, the launch flip-flop FF1 is triggered first and then the destination flip-flop is triggered after buffer delay. The skew is called as positive clock skew.

Negative clock Skew

If the capture flip-flop is triggered first and then the launch flip-flop then the design has negative clock skew. As shown in Fig. 8.2, the capture flip-flop FF0 is triggered first and then after the buffer delay the launch flip-flop FF1 is triggered and hence skew id negative clock skew.

8.2 Frequency Calculations

As discussed in the previous chapters, we have four timing paths

1. Input to reg path
2. Reg to output path
3. Reg to reg path

V. Taraate, *Advanced Digital Design Techniques*,
https://doi.org/10.1007/978-981-95-4510-0_8

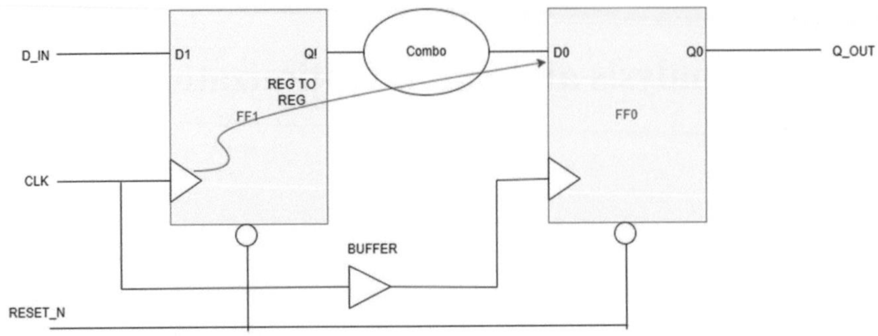

Fig. 8.1 Positive clock skew

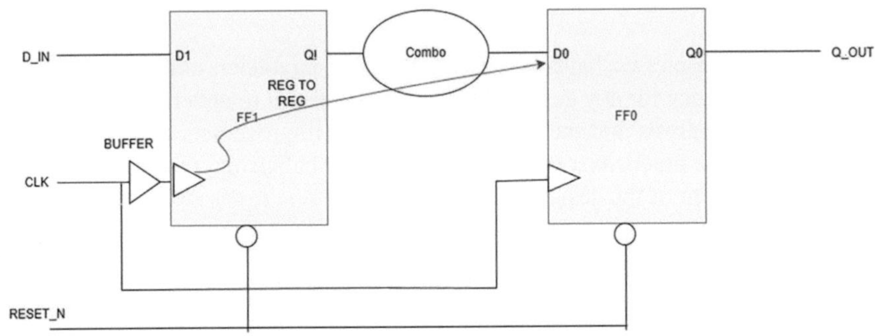

Fig. 8.2 Negative clock skew

4. Input to output unconstrained combo path.

Let us consider the design having reg to reg path and to find the maximum frequency of the design follow the following steps

1. Find the RT which is required time
   a. RT = TCLK − tsu
2. Find the AT that is arrival time
   a. AT = tctoq1 + tcombo
3. Find setup slack
   a. Setup slack = RT − AT
   b. The slack should be positive to meet timing
4. Find the TCLK
   a. RT − AT = 0
   b. TCLK − tsu − (tctoq1 + tcomb0) = 0
   c. TCLK = tctoq1 + tcombo + tsu

5. Find fmax = 1/TCLK

8.2.1 Design I

In this design (Fig. 8.3) we have three timing paths

1. D_IN to the D1
2. CLK to Q_OUT
3. Reg to reg path: start point is CLK of FF1 and end point if D0 of FF0.

Let us consider the reg-to-reg path as we have combo logic between the output of FF1 and input of FF0.

Consider both the flip-flops have tsu = 0.5 ns, th = 0.25 ns, tcombo = 1 ns, tctoq1 = tctoq2 = 1 ns.

1. Find the RT which is required time
   a. RT = TCLK − tsu = TCLK − 0.5 ns
2. Find the AT that is arrival time
   a. AT = tctoq1 + tcombo = 1 ns + 1 ns = 2 ns
3. Find setup slack
   a. Setup slack = RT − AT = TCLK − 0.5–2 ns = TCLK − 2.5 ns
   b. The slack should be positive to meet timing
4. Find the TCLK
   a. RT − AT = 0
   b. TCLK − tsu − (tctoq1 + tcomb0) = 0
   c. TCLK = tctoq1 + tcombo + tsu
   d. TCLK = 2.5 ns
5. Find fmax = 1/TCLK
6. Fmax = 400 MHz (Fig. 8.3).

8.2.2 Design II

Consider both the flip-flops have tsu = 0.5 ns, th = 0.25 ns, tcombo = 1 ns, tctoq1 = tctoq2 = 1 ns.

1. Find the RT which is required time
   a. RT = TCLK − tsu = TCLK − 0.5 ns
2. Find the AT that is arrival time (output of the flip-flop is fed back to the input)

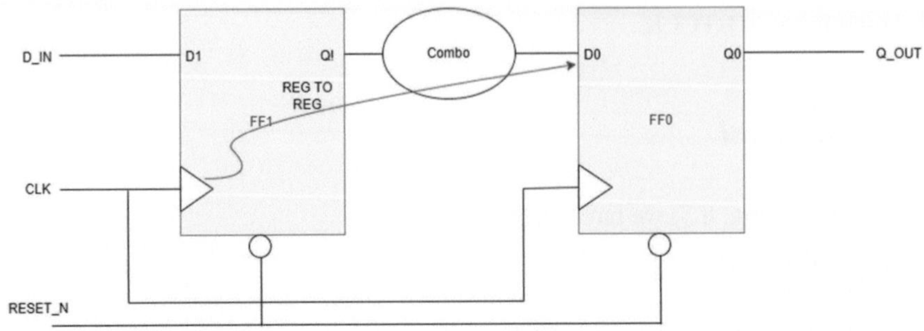

Fig. 8.3 The reg to reg path

a. AT = tctoq1 = 1 ns = 1 ns

3. Find setup slack

   a. Setup slack = RT − AT = TCLK − 0.5–1 ns = TCLK − 1.5 ns
   b. The slack should be positive to meet timing

4. Find the TCLK

   a. RT − AT = 0
   b. TCLK − tsu − (tctoq1 + tcomb0) = 0
   c. TCLK = tctoq1 + tcombo + tsu
   d. TCLK = 1.5 ns

5. Find fmax = 1/TCLK
6. Fmax = 666.66 MHz (Fig. 8.4).

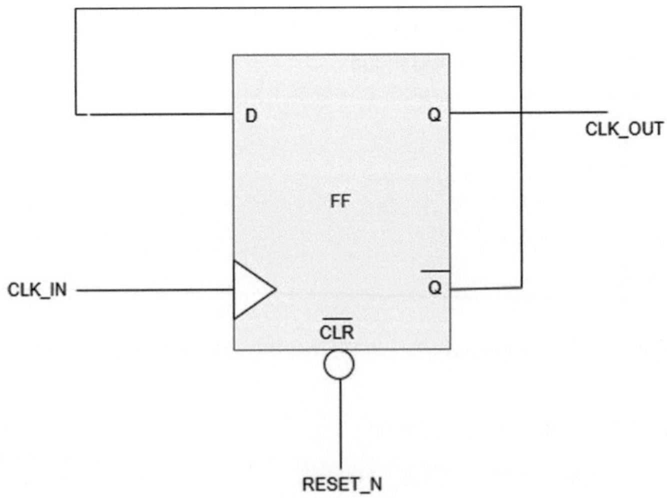

Fig. 8.4 The clock divider logic

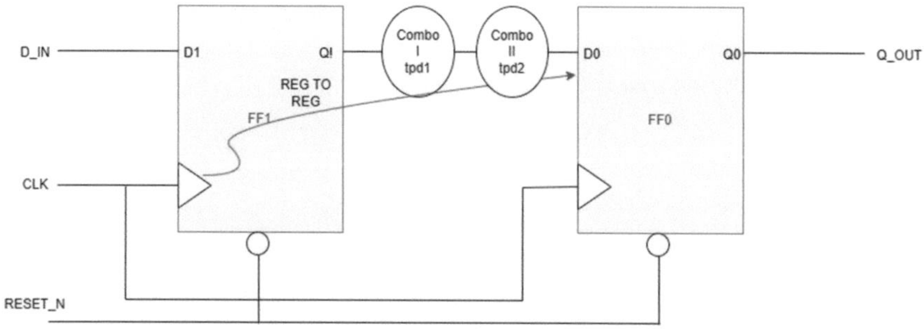

Fig. 8.5 The reg-to-reg path

8.2.3 *Design III*

Consider both the flip-flops have tsu = 0.5 ns, th = 0.25 ns, tcombo1 = 1 ns, tcombo2 = 0.5 ns, tctoq1 = tctoq2 = 1 ns.

1. Find the RT which is required time
   a. RT = TCLK − tsu = TCLK − 0.5 ns
2. Find the AT that is arrival time
   a. AT = tctoq1 + tcombo1 + tcombo2 = 1 ns + 1 ns + 0.5 ns = 2.5 ns
3. Find setup slack
   a. Setup slack = RT − AT = TCLK − 0.5–2.5 ns = TCLK − 3.0 ns
   b. The slack should be positive to meet timing
4. Find the TCLK
   a. RT − AT = 0
   b. TCLK − tsu − (tctoq1 + tcomb0) = 0
   c. TCLK = tctoq1 + tcombo + tsu
   d. TCLK = 3.0 ns
5. Find fmax = 1/TCLK
6. Fmax = 333.33 MHz (Fig. 8.5).

8.3 Skew in the Design and Impact on the Frequency

Consider the design has positive skew and shown in Fig. 8.6, to find the max frequency use following parameters

Consider both the flip-flops have tsu = 0.5 ns, th = 0.25 ns, tcombo = 1 ns, tctoq1 = tctoq2 = 1 ns, tbuf = 1 ns.

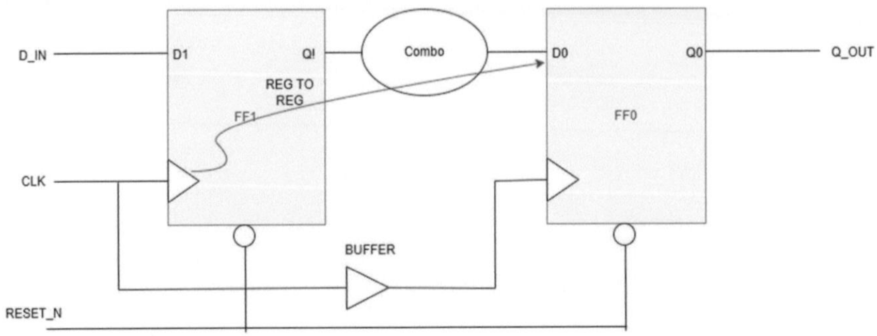

Fig. 8.6 The design having positive skew

1. Find the RT which is required time
   a. RT = TCLK + tbuf – tsu = TCLK + 1–0.5 ns = TCLK + 0.5 ns
2. Find the AT that is arrival time
   a. AT = tctoq1 + tcombo = 1 ns + 1 ns = 2 ns
3. Find setup slack
   a. Setup slack = RT – AT = TCLK + 0.5–2 ns = TCLK – 1.5 ns
   b. The slack should be positive to meet timing
4. Find the TCLK
   a. RT – AT = 0
   b. TCLK – tsu – (tctoq1 + tcomb0) = 0
   c. TCLK = tctoq1 + tcombo + tsu
   d. TCLK = 1.5 ns
5. Find fmax = 1/TCLK
6. Fmax = 666.66 MHz

Thus, the positive clock skew is useful to increase the overall frequency of the design.

Consider the design has negative clock skew and shown in Fig. 8.7 to find the max frequency use following parameters

Consider both the flip-flops have tsu = 0.5 ns, th = 0.25 ns, tcombo = 1 ns, tctoq1 = tctoq2 = 1 ns, tbuf = 1 ns.

1. Find the RT which is required time
   a. RT = TCLK – tsu = TCLK – 0.5 ns = TCLK – 0.5 ns
2. Find the AT that is arrival time

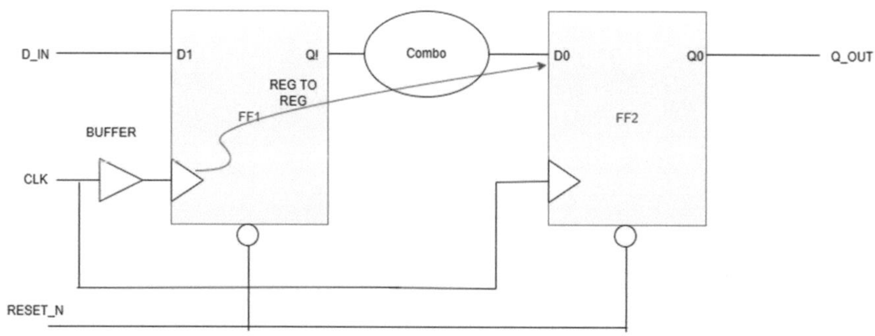

Fig. 8.7 The design having negative clock skew

   a. AT = tbuf + tctoq1 + tcombo = 1 ns + 1 ns + 1 ns = 3 ns

3. Find setup slack

   a. Setup slack = RT − AT = TCLK − 0.5–3 ns = TCLK − 3.5 ns
   b. The slack should be positive to meet timing

4. Find the TCLK

   a. RT − AT = 0
   b. TCLK − tsu − (tctoq1 + tcomb0) = 0
   c. TCLK = tctoq1 + tcombo + tsu
   d. TCLK = 3.5 ns

5. Find fmax = 1/TCLK
6. Fmax = 285.71 MHz

Thus, the negative clock skew decreases the clock frequency as source flip-flop is triggered last and destination flip-flop is triggered first.

8.4 Timing Optimization

In Sect. 8.2.3 example we have seen the cascade combo logic in the reg-to-reg path and it limits the maximum frequency of the design.

To improve the design performance let us use the register balancing technique by inserting the flop in the design. The design with the register balancing is shown in Fig. 8.8

Consider both the flip-flops have tsu = 0.5 ns, th = 0.25 ns, tcombo1 = 1 ns, tcombo2 = 0.5 ns, tctoq1 = tctoq2 = 1 ns.

(a) ***The reg to reg path I having the tcombo1 = 1 ns***

1. Find the RT which is required time

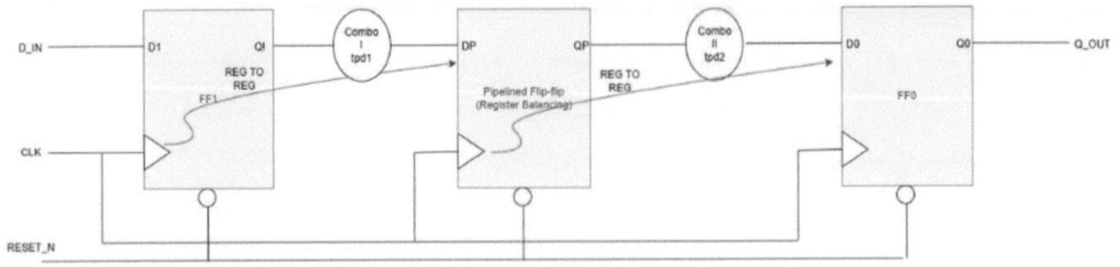

Fig. 8.8 The register balancing

   a. RT = TCLK – tsu = TCLK – 0.5 ns

2. Find the AT that is arrival time
   a. AT = tctoq1 + tcombo1 = 1 ns + 1 ns = 2.0 ns
3. Find setup slack
   a. Setup slack = RT – AT = TCLK – 0.5–2.0 ns = TCLK – 2.5 ns
   b. The slack should be positive to meet timing
4. Find the TCLK
   a. RT – AT = 0
   b. TCLK – tsu – (tctoq1 + tcomb0) = 0
   c. TCLK = tctoq1 + tcombo + tsu
   d. TCLK = 2.5 ns
5. Find fmax = 1/TCLK
6. Fmax = 400.00 MHz

(b) ***The reg to reg path I having the tcombo2 = 0.5 ns***

1. Find the RT which is required time
   a. RT = TCLK – tsu = TCLK – 0.5 ns
2. Find the AT that is arrival time
   a. AT = tctoq1 + tcombo1 = 1 + 0.5 ns = 1.5 ns
3. Find setup slack
   a. Setup slack = RT – AT = TCLK – 0.5–1.5 ns = TCLK – 2.0 ns
   b. The slack should be positive to meet timing
4. Find the TCLK
   a. RT – AT = 0
   b. TCLK – tsu – (tctoq1 + tcomb0) = 0
   c. TCLK = tctoq1 + tcombo + tsu
   d. TCLK = 2.0 ns

5. Find fmax = 1/TCLK
6. Fmax = 500.00 MHz

c) ***The max frequency of the design***

Fmax = min(400 MHz, 500 MHz) = 400 MHz the reason being if we use the 500 MHz then the reg to reg timing path 1 will have the timing violation. Hence the max frequency at which there will not be any timing violation is equal to 400 MHz.

8.5 Pipelining

In the processor designs we need to use the pipelined stages to perform the operations. For example, if we have RISCV processor then we have the five pipeline stages.

1. Fetch
2. Decode
3. Execute
4. Write-back
5. Store.

Without pipelined logic if we need to have the five clock cycles to perform the single operation then for the five instructions, we need to have 25 cycle clock consumption.

If we use the pipelined logic shown below then it minimizes the cycle consumption as the instructions are in pipeline stages.

The four stage pipelined logic stages are shown in Fig. 8.9.

The pipeline of the instruction is shown in Table 8.1.

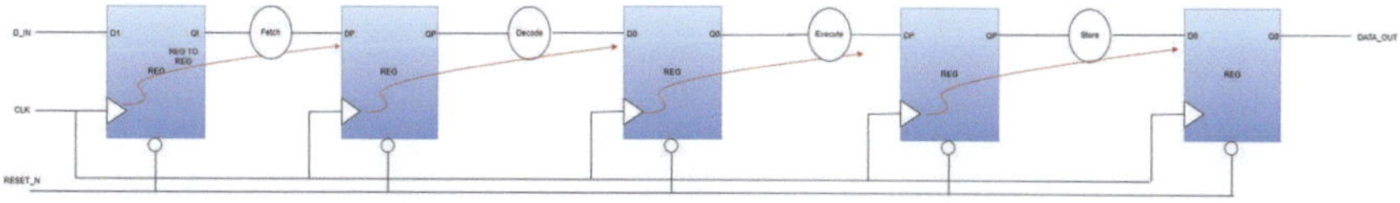

Fig. 8.9 The pipelined stages in the processor

Table 8.1 Pipelined table

Instruction	Fetch	Decode	Execute	Store
ADD	ADD	X	X	X
SUB	SUB	ADD	X	X
INC	INC	SUB	ADD	X
DEC	DEC	INC	SUB	ADD
XOR	XOR	DEC	INC	SUB

As shown in Table 8.1, the design with pipeline minimizes the cycle consumption. To perform the four instructions execution, it uses just 7 clock cycles instead of 16 clock cycles and hence saving of 9 clock cycles.

8.6 Important Highlights

Following are few of the important points to summarize this chapter.

1. The positive clock skew increases the max operating frequency of the design.
2. The negative clock skew reduces the frequency of the design.
3. The launch flip-flop is triggered first and capture flip-flop is triggered last is called as positive clock skew.
4. The launch flip-flop is triggered last and capture flip-flop is triggered first is called as negative clock skew.
5. The register balancing is used to improve the design performance.
6. The pipeline logic reduces the cycle consumption.

8.7 Puzzle

Consider the 2-bit gray counter and find the maximum operating frequency of the design. Consider timing parameters as tcombo = 1 ns, tctoq = 1 ns, tsu = 0.5 ns and th = 0.25 ns.

Chapter 9
Synchronizers and High Speed Designs

The various synchronizers which are used in the multiple clock domains are discussed in this chapter. The chapter covers the control path, data path synchronizers and practical considerations and scenarios during the design. The chapter also covers the reset synchronizers and the strategies for the deassertion and reset domain crossing issues.

9.1 Level Synchronizers

Consider the multiple clock domain designs and control signals need to be passed from the clock domain I to the clock domain II. In such scenario we need to deploy the level synchronizers in the design.

The two flop level synchronizer is shown in Fig. 9.1 and used to get the valid data output Q_OUT on the second clock edge. The FF1 data output is metastable and during the timing analysis we can disable the timing of the FF1.

To avoid the corrupt data due to metastability we can use the 2-flop or three flop level synchronizers.

The timing diagram is shown in Fig. 9.2.

9.2 Multi-bit Signal Passing

As discussed in the above section, the one bit signal passing from the clock domain I to the clock domain II can be achieved successfully by the level synchronizer. For the multi-bit signal passing from the clock domain I to the clock domain II can be achieved by the gray codes. In the binary codes one or more than one-bit changes in two consecutive binary numbers. In the gray codes it is guaranteed to have the

V. Taraate, *Advanced Digital Design Techniques*,
https://doi.org/10.1007/978-981-95-4510-0_9

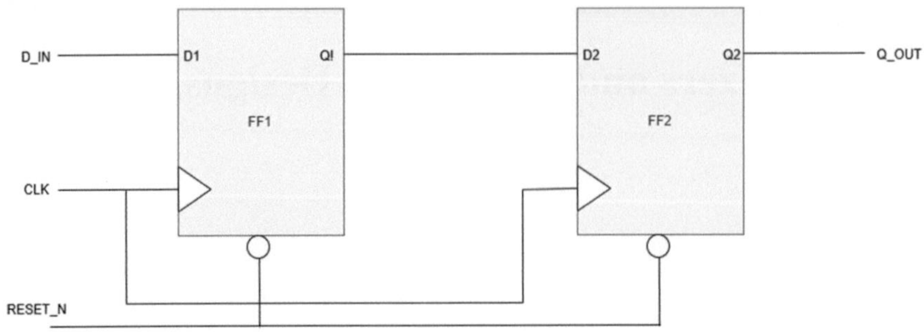

Fig. 9.1 The two-flop synchronizer

Fig. 9.2 The timing diagram of the two-flop synchronizer

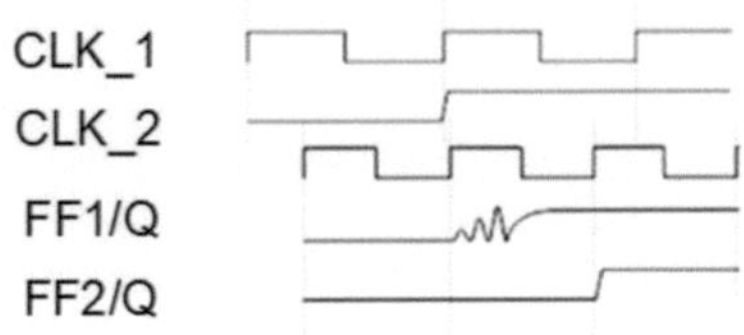

single bit change and this avoids the issue due to metastability. The design is shown in Fig. 9.3.

As shown, the 4-bit binary number is converted into the gray code in the clock domain I and registered output from clock domain I is passed to the clock domain II using the multi-bit level synchronizers. At the clock domain II the binary value is recovered using the gray to binary code converter.

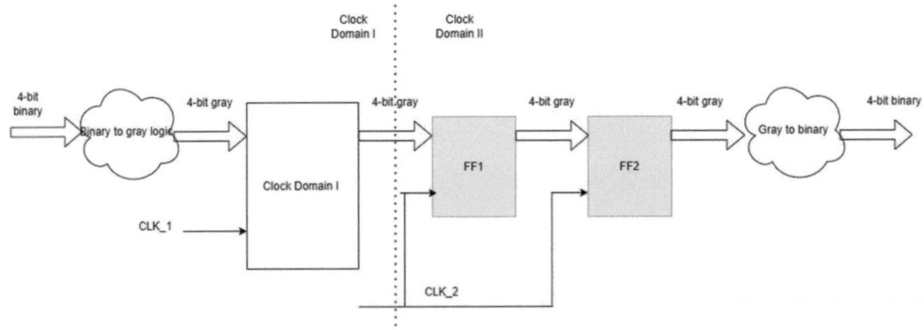

Fig. 9.3 The gray encoding for the CDC

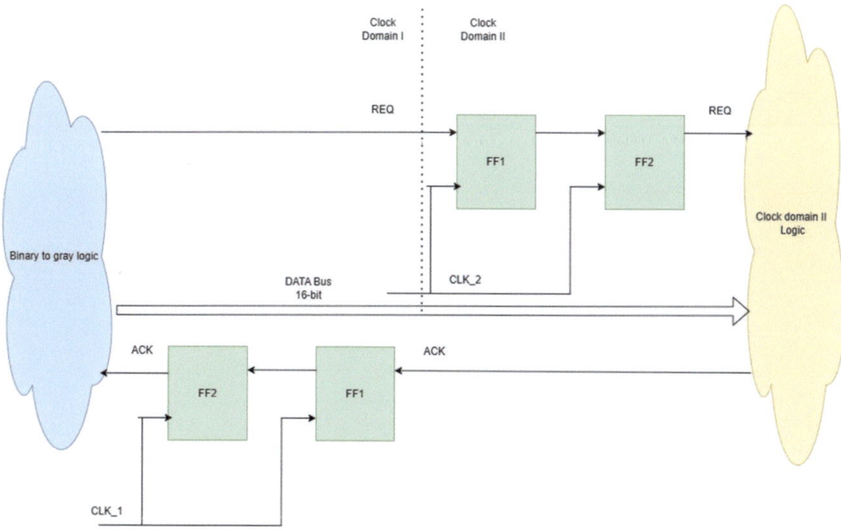

Fig. 9.4 The handshake synchronization concept

9.3 Handshake Synchronizers

The handshake synchronizers using the request (REQ) and acknowledge (ACK) can be used to pass the multi-bit data from the clock domain I to the clock domain II. The handshake concept is shown in Fig. 9.4.

The synchronization mechanism uses the level synchronizers to pass the REQ from the clock domain I to the clock domain II. The ACK from the clock domain II is passed through the level synchronizer and used by the clock domain I logic.

The main application in the FSM based design where the clock domain I FSM controller can generate the valid request (REQ) to send the data and the clock domain II FSM controller generates the acknowledgement (ACK) signal after receiving the data. The synchronization mechanism is shown in Fig. 9.5.

This kind of the synchronization technique is useful if the data is stable and doesn't change frequently unless and until the ack is received.

9.4 Mux Synchronizer

They are used to isolate the multi-bit data as shown in Fig. 9.6. The clock domain I generate the control signal when the data is available in the clock domain I and this control pulse is synchronized between the two clock domains using the level synchronizer or the pulse synchronizer depending on the write to read clock ratio.

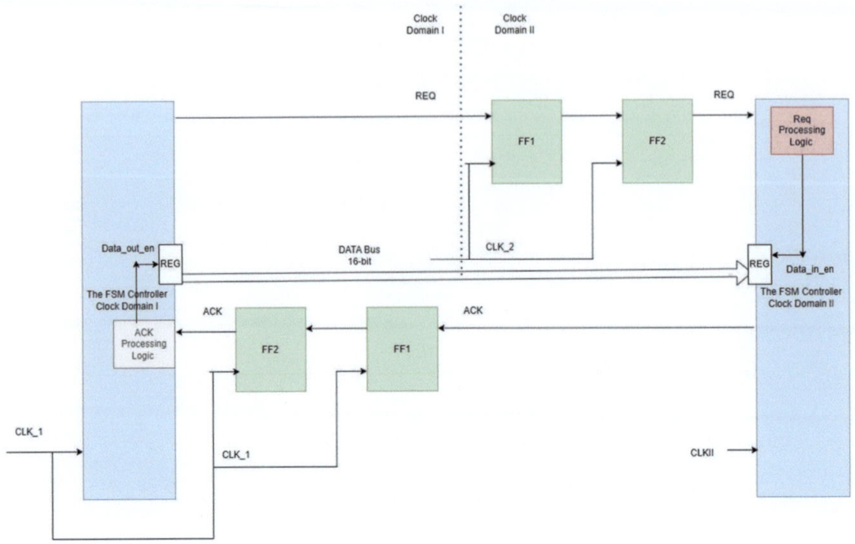

Fig. 9.5 FSM data synchronization

And then used as a control pulse in the clock domain II. The data is stable until it is sampled using the synchronized control pulse.

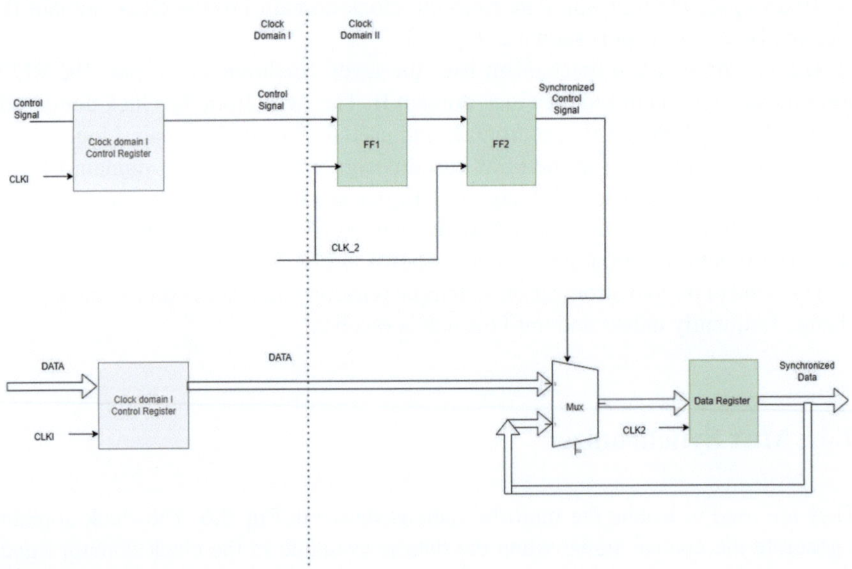

Fig. 9.6 The mux synchronizer

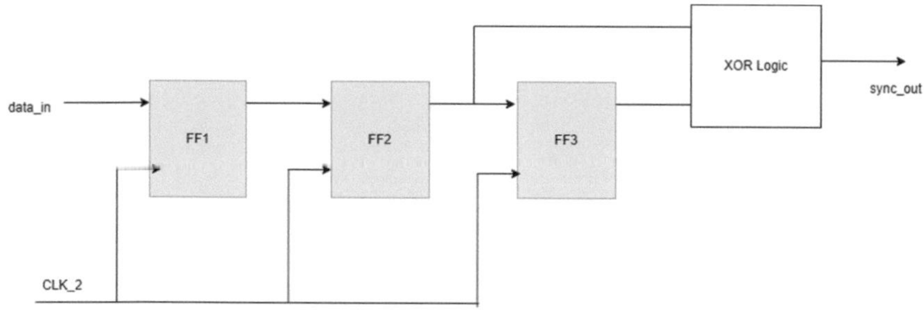

Fig. 9.7 The pulse synchronizer

9.5 Pulse Synchronizer

If the control signal changes very fast then we can use the pulse synchronizer. The pulse synchronizer is used to sample the single clock wide pulse from the clock domain I and to generate the single clock wide pulse in the clock domain II.

The design of the pulse synchronizer is shown in Fig. 9.7.

9.6 Reset Synchronizers

Most of the time we have the asynchronous reset in the design and reset Can be asserted at any instance of time as it is asynchronous in the nature. If we consider the reset recovery and reset removal time then during the deassertion there are issues and hence we need to have the reset synchronizers. The assertion of the reset is asynchronous and deassertion is synchronous.

Following are few of the important points we need to consider while designing the reset synchronizers

1. Consider the reset recovery and reset removal time.
2. The reset synchronizer is used to have synchronous deassertion.
3. The reset synchronizer must fanout to all the SoC registers that need to be out of the reset in a single cycle. And there must be a single synchronizer for all such flops. Some flops will be out of reset n cycles later and need to consider the cycle count required for the deassertion of the reset to come out of the reset phase.
4. If the SoC has the multiple clock domains then those many reset synchronizers need to be used and deployed in the respective clock domain.

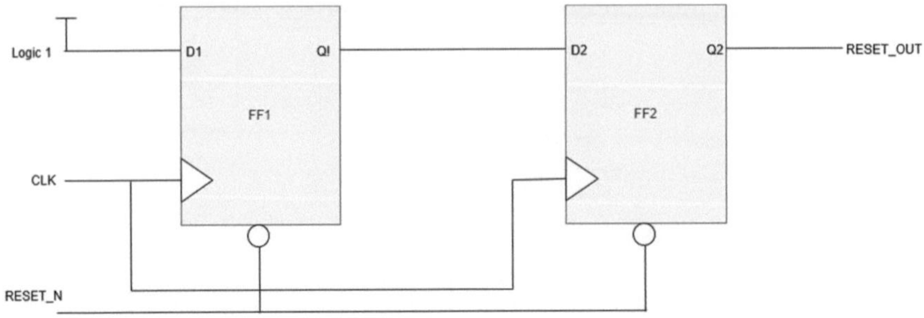

Fig. 9.8 The reset synchronizer (active low reset_n)

9.6.1 Active Low Asynchronous Reset

If the reset is active low asynchronous reset, then during the normal operation reset should be logic 1 and we can deploy the following synchronizer for the synchronous deassertion. The asynchronous assertion is not a problem here. When the reset is released then the first flip-flop output is forced as logic 1 and on the second clock cycle the RESET_OUT is equal to 1 to indicate the completion of the reset deassertion phase (Fig. 9.8).

9.6.2 Active High Asynchronous Reset

If the reset is active high asynchronous reset, then during the normal operation reset should be logic 0 and we can deploy the following synchronizer for the synchronous deassertion. The asynchronous assertion is not a problem here. When the reset is released then the first flip-flop output is forced as logic 0 and on the second clock cycle the RESET_OUT is equal to 0 to indicate the completion of the reset deassertion phase (Fig. 9.9).

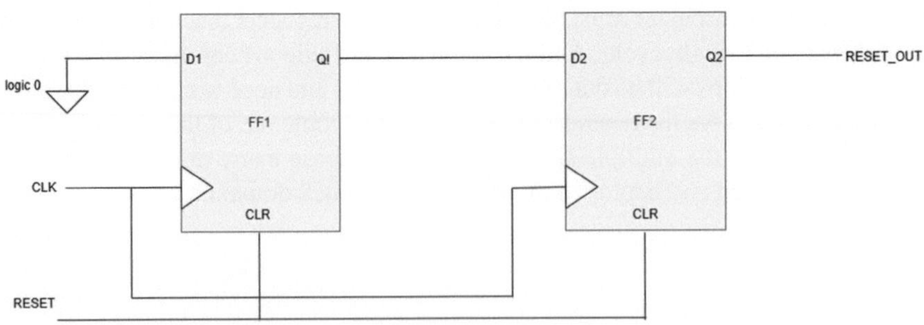

Fig. 9.9 The reset synchronizer (active high reset)

9.7 Data Path Synchronizers

In the previous section we have discussed the control path synchronizers and the reset synchronizers. In this section let us discuss the data path synchronizers that is FIFOs.

The asynchronous FIFO architecture is shown in Fig. 9.10.

The design has the following components

1. Dual port RAM
2. Write Control Logic
3. Read Control Logic
4. Synchronizers for the pointer comparison
5. The FIFO full and empty logic.

Depending on the write, read clock the FIFO depth can vary and following section is useful to understand few of the practical scenarios to get the depth of the FIFO.

9.7.1 *Faster to Slower Clock Domain*

Consider the design having write clk frequency f1 = 200 MHz and read clock frequency f2 = 100 MHz and the burst length is 80 byte find the depth of FIFO?

Solution: To find the depth of the FIFO use the following steps.

1. **Time required to write single data byte** (T_{W})

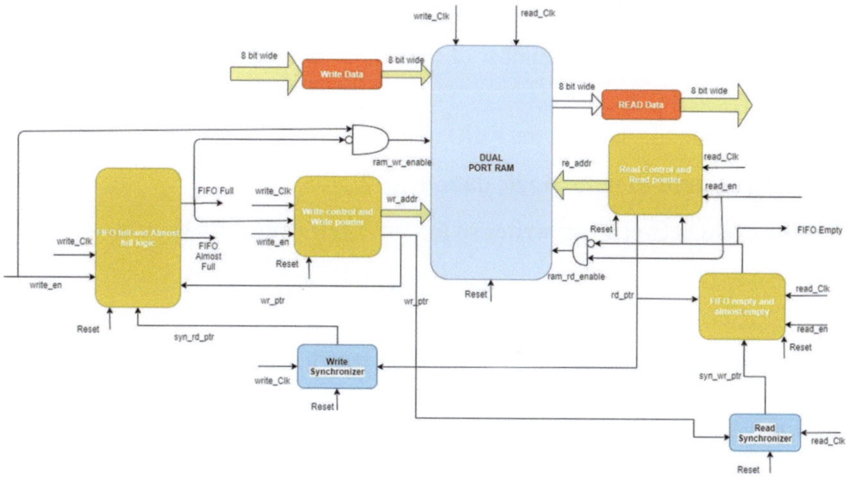

Fig. 9.10 The asynchronous FIFO architecture

$$T_w = 1/200\,\text{MHz} = 5\,\text{ns}$$

2. **Time required to write burst of the data that is 40 bytes** (T_{b_w})

$$T_{b\_w} = T_w * \text{Burst length} = 5\,\text{ns} * 80 = 400\,\text{ns}$$

3. **Time required to read one data** (T_r)

$$T_r = 1/100\,\text{MHz} = 10\,\text{ns}$$

4. **The number of data reads in duration of** T_{b_w}

$$\text{No of reads} = 400\,\text{ns}/10\,\text{ns} = 40$$

5. **The depth of FIFO**

$$\begin{aligned}\text{Depth of FIFO} &= \text{Burst length} - \text{No of reads}\\ &= 80 - 40 = 40\end{aligned}$$

9.7.2 *Faster to Slower Clock Domain with Idle Cycles*

Consider the design having write clk frequency f1 = 100 MHz and read clock frequency f2 = 50 MHz and the burst length is 40 bytes. The number of idle cycles between two writes is 1 and number of idle cycle between two reads is 3. Find the depth of FIFO?

Solution: To find the depth of the FIFO use the following steps.

1. **Time required to write single data byte** (T_w)

One idle cycle between two writes so for two cycles one data is written.

$$T_w = 2 * (1/100\,\text{MHz}) = 20\,\text{ns}$$

2. **Time required to write burst of the data that is 40 bytes** (T_{b_w})

$$T_{b\_w} = T_w * \text{Burst length} = 20\,\text{ns} * 40 = 800\,\text{ns}$$

3. **Time required to read one data** (T_r)

For two successive reads three cycles (2 idle + 1 = 3 cycles)

$$T_r = 4 * (1/50\,\text{MHz}) = 80\,\text{ns}$$

4. **The number of data reads in duration of** T_{b_w}

$$\text{No of reads} = 800\,\text{ns}/80\,\text{ns} = 10$$

5. **The depth of FIFO**

$$\begin{aligned}\text{Depth of FIFO} &= \text{Burst length} - \text{No of reads}\\ &= 40 - 10 = 30.\end{aligned}$$

9.7.3 *Slower to Faster Clock Domains*

Consider the design having write clk frequency f1 = 50 MHz and read clock frequency f2 = 100 MHz and the burst length is 100 bytes. The number of idle cycles between two writes is 1 and number of idle cycle between two reads is 3.Find the depth of FIFO?

Solution: To find the depth of the FIFO use the following steps.

1. **Time required to write single data byte** (T_w)

One idle cycle between two writes so for two cycles one data is written.

$$T_w = 2^*(1/50\,\text{MHz}) = 40\,\text{ns}$$

2. **Time required to write burst of the data that is 100 bytes** (T_{b_w})

$$T_{b\_w} = T_w * \text{Burst length} = 40\,\text{ns} * 100 = 4000\,\text{ns}$$

3. **Time required to read one data** (T_r)

For two successive reads three cycles (2 idle + 1 = 3 cycles)

$$T_r = 4 * (1/100\,\text{MHz}) = 40\,\text{ns}$$

4. **The number of data reads in duration of** T_{b_w}

$$\text{No of reads} = 4000\,\text{ns}/40\,\text{ns} = 100$$

5. **The depth of FIFO**

$$\text{Depth of FIFO} = \text{Burst length} - \text{No of reads}$$
$$= 100 - 100 = 0\ \text{NO FIFO required}$$

9.7.4 Slower to Faster Clock Domain with Idle Cycle

Consider the design having write clk frequency f1 = 25 MHz and read clock frequency f2 = 40 MHz and the burst length is 100 bytes. The number of idle cycles between two writes is 1 and number of idle cycle between two reads is 3.Find the depth of FIFO?

Solution: To find the depth of the FIFO use the following steps.

1. **Time required to write single data byte** (T_w)

One idle cycle between two writes so for two cycles one data is written.

$$T_w = 2 * (1/25\,\text{MHz}) = 80\,\text{ns}$$

2. **Time required to write burst of the data that is 100 bytes** (T_{b_w})

$$T_{b\_w} = T_w * \text{Burst length} = 80\,\text{ns} * 100 = 8000\,\text{ns}$$

3. **Time required to read one data** (T_r)

For two successive reads three cycles (2 idle + 1 = 3 cycles)

$$T_r = 4 * (1/40\,\text{MHz}) = 100\,\text{ns}$$

4. **The number of data reads in duration of** T_{b_w}

$$\text{No of reads} = 8000\,\text{ns}/100\,\text{ns} = 80$$

5. **The depth of FIFO**

$$\text{Depth of FIFO} = \text{Burst length} - \text{No of reads}$$
$$= 100 - 80 = 20$$

9.7.5 *Same Frequency in Both Clock Domain*

Consider the design having write clk frequency f1 = 50 MHz and read clock frequency f2 = 50 MHz and the burst length is 80 bytes. The number of idle cycles between two writes is 1 and number of idle cycle between two reads is 3.Find the depth of FIFO?

Solution: To find the depth of the FIFO use the following steps.

1. **Time required to write single data byte** (T_w)

One idle cycle between two writes so for two cycles one data is written.

$$T_w = 2 * (1/50\,\text{MHz}) = 40\,\text{ns}$$

2. **Time required to write burst of the data that is 80 bytes** (T_{b_w})

$$T_{b\_w} = T_w * \text{Burst length} = 40\,\text{ns} * 80 = 3200\,\text{ns}$$

3. **Time required to read one data** (T_r)

For two successive reads three cycles (2 idle + 1 = 3 cycles)

$$T_r = 4 * (\,1/50\,\text{MHz}) = 80\,\text{ns}$$

4. **The number of data reads in duration of** T_{b_w}

$$\text{No of reads} = 3200\,\text{ns}/80\,\text{ns} = 40$$

5. **The depth of FIFO**

$$\begin{aligned}\text{Depth of FIFO} &= \text{Burst length} - \text{No of reads}\\ &= 80 - 40 = 40.\end{aligned}$$

9.8 Important Highlights

Following are few of the important points to summarize this chapter.

1. The synchronizers are used in the multiple clock domain designs in the control path and data paths
2. The level synchronizers using the two-flop synchronizer are used in the control path and useful to pass the single bit control signal from clock domain I to clock domain 2

3. The gray encoding techniques are used to pass the multi-bit signals from clock domain I to clock domain 2
4. The handshake mechanism can be used to pass the multi-bit data from clock domain I to clock domain 2
5. The reset synchronizers are used to resolve the issues during the deassertion of the reset.
6. The FIFOs are used as the data path synchronizers in the high speed designs.

9.9 Puzzle

Consider the design has the three clock domains clk domain 1 = 500 MHz, the clock domain 2 = 250 MHz and clock domain 3 = 333.33 MHz sketch the top-level details to pass the control and data signal in these clock domains.

Chapter 10
FSM and FSMD

The FSM are finite state machines and used to design the FSM controllers. For example, to detect the sequence of 1010 from the input we can think of FSM designs. The arbitrary counters, sequence detectors, controllers can be designed using efficient FSM design techniques. In the previous chapters we have discussed about the combinational and sequential design techniques. In this chapter let us discuss about the FSM design techniques and their applications in the digital design.

10.1 FSM Designs

FSM designs are Finite State machines and they are used in the SoC design to generate the control and timing signals.

The FSMs are classified mainly into two categories

1. Moore FSM
2. Mealy FSM

Now let us discuss the difference between the Moore and Mealy FSMs and their use in the SoC designs.

10.1.1 Moore FSM

In the Moore FSM the output is function of the current state only and not dependent on the changes in the input. As output is function of the current state the Moore machine needs one extra state as compared to the Mealy machine.

V. Taraate, *Advanced Digital Design Techniques*,
https://doi.org/10.1007/978-981-95-4510-0_10

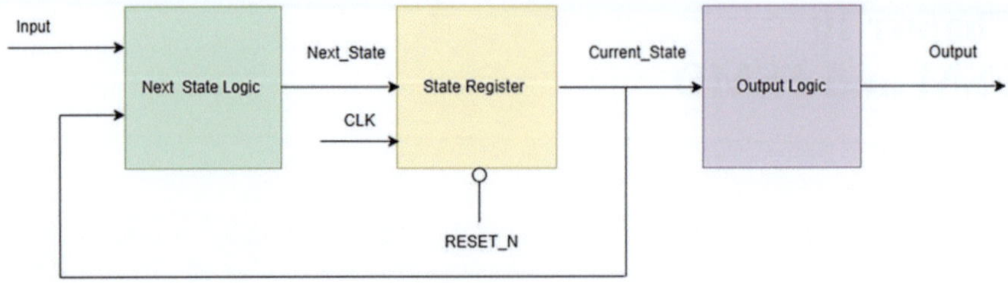

Fig. 10.1 Architecture diagram of Moore machine

The Moore FSM is having the following functional blocks

1. State register: Sequential logic
2. Next state logic: Combinational Logic
3. Output logic: Combinational Logic

The architecture of the Moore FSM is shown in Fig. 10.1.

10.1.2 Mealy FSM

In the Mealy FSM design output is function of the current state and input. The Mealy machine needs one state less as compared to the Moore machine and faster as compared to Moore machine.

The Mealy FSM is designed using the following blocks

1. State register: Sequential Logic
2. Next state logic: The combinational Logic
3. Output logic: Combinational Logic

The architecture diagram of the Mealy FSM is shown in Fig. 10.2.

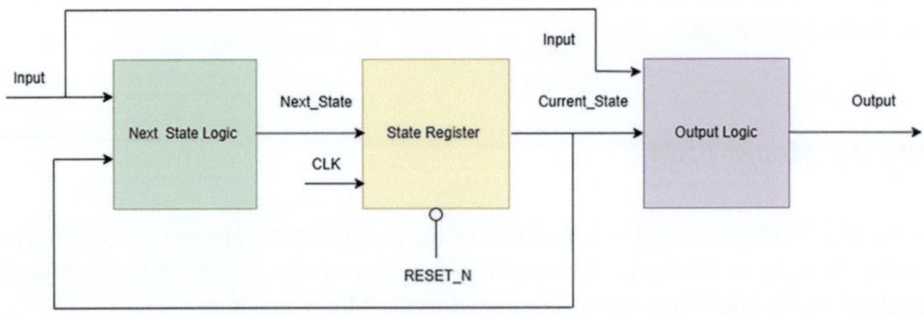

Fig. 10.2 Architecture diagram of Mealy machine

Table 10.1 Differences between Moore and Mealy machines

Moore machine	Mealy machine
Outputs are function of present state only	Outputs are function of the present state and inputs also
Output is stable for one clock cycle and Moore FSM are not prone to glitches or spikes	Output may change multiple times depending on changes in the input and may or may not be stable for one clock cycle hence Mealy FSM are prone to glitches or hazards
It requires a greater number of states as compared to Mealy machine	Mealy machine needs lesser states as compared to Moore machine
Moore FSM has the higher operating frequency as compared to Mealy machine	Mealy FSM has the lower operating frequency as compared to Moore machine

10.1.3 Moore Versus Mealy FSM

The difference between the Moore and Mealy FSM is documented in the Table 10.1.

10.2 State Encoding Methods

For the FSM designs we should understand the state encoding. There are three types of the state encoding for FSM based designs and they are

1. **Binary encoding**

In the binary encoding if the number of states are m, then the number of flip-flops required is computed using

$$n = \log_2 m$$

where n = number of flip-flops and m are number of states.

Consider the number of states as $m = 4$ then number of flip-flops needed to design FSM are

$$n = \log_2 4$$

The four states are represented as

S0 = 00
S1 = 01
S2 = 10
S3 = 11

2. **Gray encoding**

In the gray encoding if the number of states are m, then the number of flip-flops required is computed using

$$n = \log_2 m$$

where $n =$ number of flip-flops and m are number of states.

Consider the number of states as $m = 4$ then number of flip-flops needed to design FSM are

$$n = \log_2 4 = 2.$$

The four states are represented as

S0 = 00
S1 = 01
S2 = 11
S3 = 10

The gray encoding are useful in the FSM design to save the power as in two successive gray numbers only one-bit changes.

3. **One-hot encoding**

In the one-hot encoding if the number of states are m, then the number of flip-flops required is computed using $n = m$. In this encoding only one bit is active high or hot at a time.

Where $n =$ number of flip-flops and m are number of states.

Consider the number of states as $m = 4$ then number of flip-flops needed to design FSM are $n = m = 4$.

The four states using one-hot encoding are represented as

S0 = 0001
S1 = 0010
S2 = 0100
S3 = 1000

The one-hot encoding are useful in the FSM design to have better timing if area is not the constraint.

10.3 Moore FSM Design

Now let us design the FSM for the given specifications. What we need to do is that we need to design the digital logic for the

1. State register
2. Next state logic
3. Output logic

Let us consider the design of the sequential circuit to detect the sequence 101 non-overlapping

1. **Find the number of states to detect the non-overlapping sequence 101 using Moore machine**

Number of states = 4. The states are S0, S1, S2, S3.

2. **State diagram**

The state transition happens as shown in Fig. 10.3.

3. **Find number of flip-flops**

We will use the binary encoding and the number of flip-flops = $n = \log_2 4 = 2$. We will use the positive edge sensitive D flip-flops.

4. **Reset strategy**

Let us use active low asynchronous reset input reset_n. For reset_$n = 0$ the counter holds the previous output. For the reset_$n = 1$ the output increments on the rising edge of the clock.

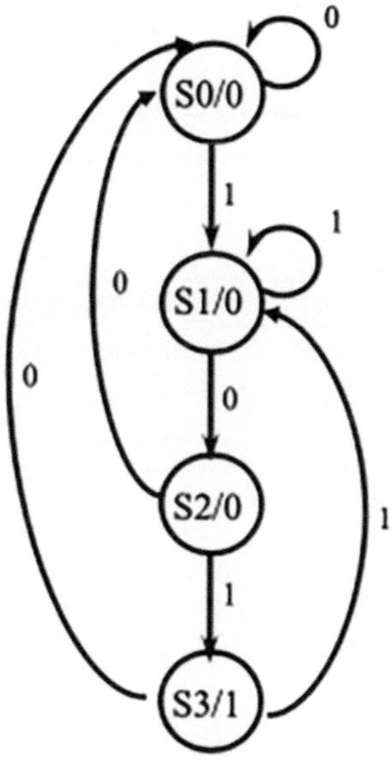

Fig. 10.3 Moore FSM of Moore 101 non-overlapping sequence detector

Table 10.2 The state table of the 101 Moore nonoverlapping sequence detector

Input (I)	Present state Q1 Q0	Next state ($Q1^+$ $Q0^+$)
0	S0 = 00	S0 = 00
1	S0 = 00	S1 = 01
0	S1 = 01	S2 = 10
1	S1 = 01	S1 = 01
0	S2 = 10	S0 = 00
1	S2 = 10	S3 = 11
0	S3 = 11	S3 = 00
1	S3 = 11	S1 = 01

5. **Let us document the entries in the state table to get state register logic**

The state table of the 101 Moore non-overlapping sequence detector is shown in Table 10.2.

So, we need to have two D flip-flop having asynchronous active low reset input reset_*n* and rising edge sensitive clock.

6. **Use of the excitation table to design the next state logic**

The excitation table consists of the information about the present state, next state, and excitation input (Table 10.3).

Now let us use the K-map to deduce the Boolean equation for the next state logic that is for the D1, D0 (Fig. 10.4).

$$D1 = \bar{I} \cdot \overline{Q1} \cdot Q0 + I \cdot \overline{Q0} \cdot Q1$$

Table 10.3 The excitation table of the 101 Moore sequence detector nonoverlapping

Input (I)	Present state Q1 Q0	Next state ($Q1^+$ $Q0^+$)	D1	D0
0	S0 = 00	S0 = 00	0	0
1	S0 = 00	S1 = 01	0	1
0	S1 = 01	S2 = 10	1	0
1	S1 = 01	S1 = 01	0	1
0	S2 = 10	S0 = 00	0	0
1	S2 = 10	S3 = 11	1	1
0	S3 = 11	S3 = 00	0	0
1	S3 = 11	S1 = 01	0	1

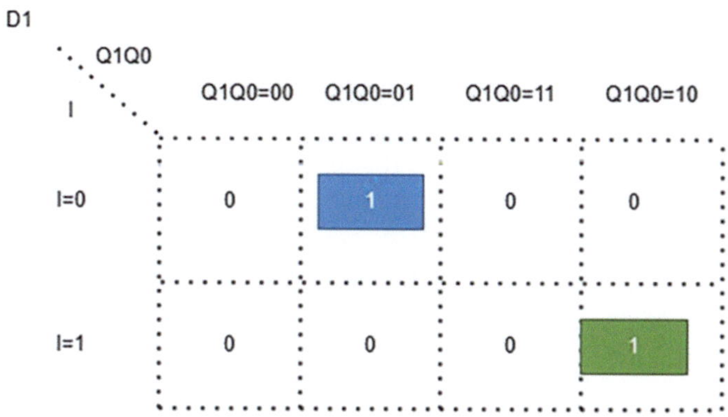

Fig. 10.4 The K-map for the next state logic

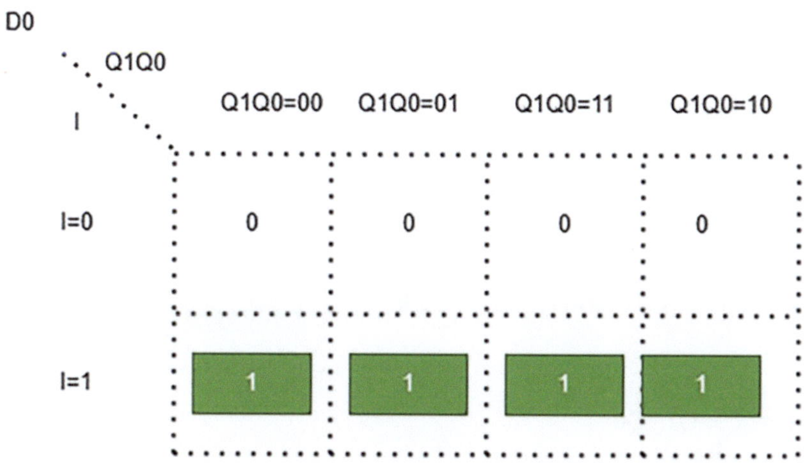

$$D0 = I$$

7. **Output logic**

In the Moore FSM output if function of the present state only (Table 10.4).

Table 10.4 The truth table of the output logic

Input (I)	Present state (Q1 Q0)	Next state ($Q1^+$ $Q0^+$)	Y
0	S0 = 00	S0 = 00	0
1	S0 = 00	S1 = 01	0
0	S1 = 01	S2 = 10	0
1	S1 = 01	S1 = 01	0
0	S2 = 10	S0 = 00	0
1	S2 = 10	S3 = 11	0
0	S3 = 11	S3 = 00	1
1	S3 = 11	S1 = 01	1

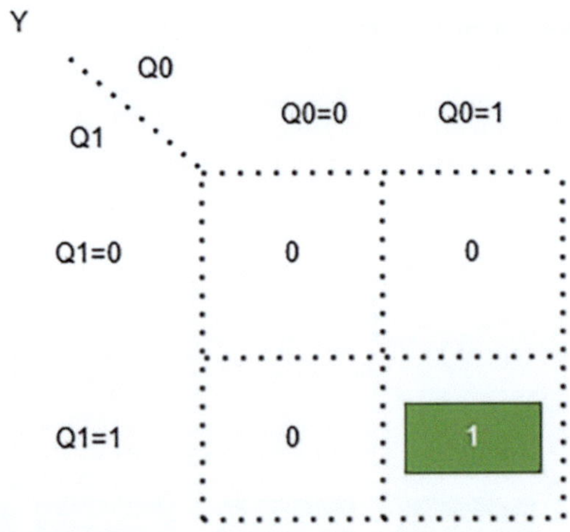

Here Y = Q1 · Q0.

8. **Let us sketch the FSM Design**

The design is ss shown in the Fig. 10.5.

10.4 Mealy FSM Design

Let us consider the design of the sequential circuit to detect the 101 non-overlapping sequence We can use the following design steps to design the Mealy FSM.

1. **Find the number of states to detect the 101 nonoverlapping Mealy sequence**

Number of states = 3. The states are S0, S1 and S2.

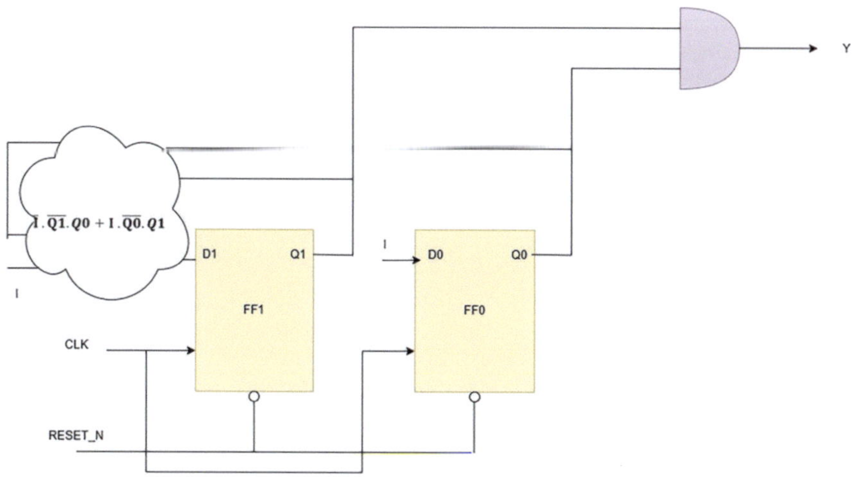

Fig. 10.5 The logic design of the 101 nonoverlapping Moore sequence detector

2. **State diagram**

The state transition happens as shown in Fig. 10.6.

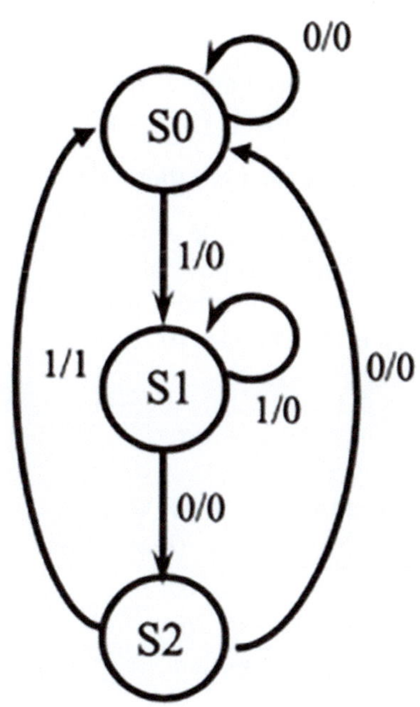

Fig. 10.6 Mealy FSM of 101 non-overlapping Mealy machine

Table 10.5 The state table of the 101 non-overlapping Mealy sequence detector

Input (I)	Present state (Q1 Q0)	Next state ($Q1^+$ $Q0^+$)
0	S0 = 00	S0 = 00
1	S0 = 00	S1 = 01
0	S1 = 01	S2 = 10
1	S1 = 01	S1 = 01
0	S2 = 10	S0 = 00
1	S2 = 10	S0 = 00

3. **Find number of flip-flops**

We will use the binary encoding and the number of flip-flops $= n = \log_2 3 = 2$ We will use the positive edge sensitive D flip-flops.

4. **Reset strategy**

Let us use active low asynchronous reset input reset_*n*. For reset_*n*=0 the sequence detector holds the default state. For the reset_*n*=1 the state transition happens on the rising edge of the clock.

5. **Let us document the entries in the state table to get state register logic**

The state table of the 101 nonoverlapping Mealy sequence detector is shown in Table 10.5.

So, we need to have two D flip-flop having asynchronous active low reset input reset_*n* and rising edge sensitive clock.

6. **Use of the excitation table to design the next state logic**

The excitation table consists of the information about the present state, next state, and excitation input (Table 10.6).

Now let us use the K-map to deduce the Boolean equation for the next state logic that is D1, D0 (Fig. 10.7).

$$D1 = \bar{I} \cdot Q0$$

Table 10.6 The excitation table of the Mealy 101 nonoverlapping sequence detector

Input (I)	Present state (Q1 Q0)	Next state ($Q1^+$ $Q0^+$)	D1	D0
0	S0 = 00	S0 = 00	0	0
1	S0 = 00	S1 = 01	0	1
0	S1 = 01	S2 = 10	1	0
1	S1 = 01	S1 = 01	0	1
0	S2 = 10	S0 = 00	0	0
1	S2 = 10	S0 = 00	0	0

D0

I \ Q1Q0	Q1Q0=00	Q1Q0=01	Q1Q0=11	Q1Q0=10
I=0	0	0	X	0
I=1	1	1	X	0

$$D0 = I \cdot \overline{Q1}$$

7. **Output logic**

In the Mealy FSM output if function of the present state and input (Table 10.7).

Now let us use the K-map to deduce the Boolean equation for the output logic (Fig. 10.8).

$$Y = I \cdot Q1$$

So, we need to have AND gate as an output logic.

D1

I \ Q1Q0	Q1Q0=00	Q1Q0=01	Q1Q0=11	Q1Q0=10
I=0	0	1	X	0
I=1	0	0	X	0

Fig. 10.7 The K-map for the next state logic

Table 10.7 The truth table of the output logic

Input (I)	Present state (Q1 Q0)	Next state ($Q1^+$ $Q0^+$)	D1
0	S0 = 00	S0 = 00	0
1	S0 = 00	S1 = 01	0
0	S1 = 01	S2 = 10	0
1	S1 = 01	S1 = 01	0
0	S2 = 10	S0 = 00	0
1	S2 = 10	S0 = 00	1

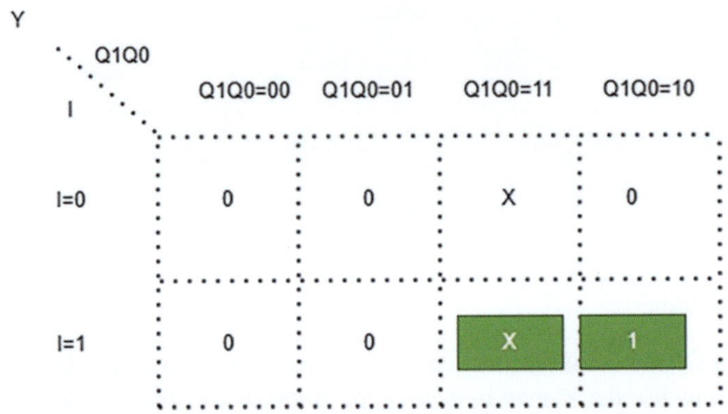

Fig. 10.8 The K-map for the output logic

8. **Let us sketch the FSM Design**

The design of the Mealy 101 nonoverlapping sequence detector is shown in Fig. 10.9.

10.5 Moore 101 Overlapping Sequence Detector Design

Now let us design the FSM for the given specifications. What we need to do is that we need to design the digital logic for the.

1. State register
2. Next state logic
3. Output logic

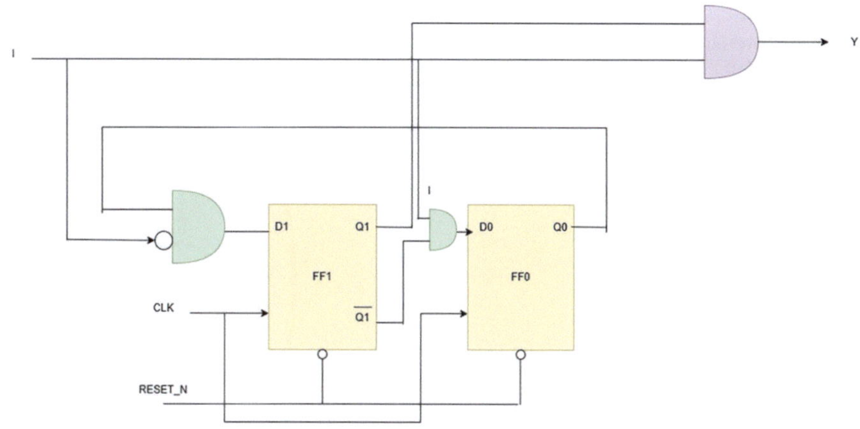

Fig. 10.9 The Mealy 101 non-overlapping sequence detector

Let us consider the design of the sequential circuit to detect the sequence 101 overlapping

1. **Find the number of states to detect the overlapping sequence 101 using Moore machine**

Number of states = 4. The states are S0, S1, S2, S3.

2. **State diagram**

The state transition happens as shown in Fig. 10.10.

3. **Find number of flip-flops**

We will use the binary encoding and the number of flip-flops $= n = \log_2 4 = 2$. We will use the positive edge sensitive D flip-flops.

4. **Reset strategy**

Let us use active low asynchronous reset input reset_*n*. For reset_$n = 0$ the counter holds the previous output. For the reset_$n = 1$ the output increments on the rising edge of the clock.

5. **Let us document the entries in the state table to get state register logic**

The state table of the 101 Moore overlapping sequence detector is shown in Table 10.8.

So, we need to have two D flip-flop having asynchronous active low reset input reset_*n* and rising edge sensitive clock.

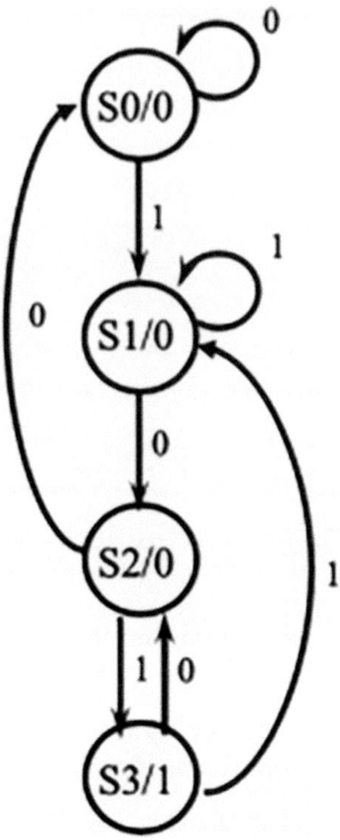

Fig. 10.10 Moore FSM of Moore 101 overlapping sequence detector

Table 10.8 The state table of the 101 Moore overlapping sequence detector

Input (I)	Present state (Q1 Q0)	Next state ($Q1^+$ $Q0^+$)
0	S0 = 00	S0 = 00
1	S0 = 00	S1 = 01
0	S1 = 01	S2 = 10
1	S1 = 01	S1 = 01
0	S2 = 10	S0 = 00
1	S2 = 10	S3 = 11
0	S3 = 11	S2 = 10
1	S3 = 11	S1 = 01

6. **Use of the excitation table to design the next state logic**

The excitation table consists of the information about the present state, next state, and excitation input (Table 10.9).

Now let us use the K-map to deduce the Boolean equation for the next state logic that is for the D1, D0 (Fig. 10.11).

Table 10.9 The excitation table of the 101 Moore sequence detector overlapping

Input (I)	Present state (Q1 Q0)	Next state ($Q1^{+}$ $Q0^{+}$)	D1	D0
0	S0 = 00	S0 = 00	0	0
1	S0 = 00	S1 = 01	0	1
0	S1 = 01	S2 = 10	1	0
1	S1 = 01	S1 = 01	0	1
0	S2 = 10	S0 = 00	0	0
1	S2 = 10	S3 = 11	1	1
0	S3 = 11	S2 = 10	1	0
1	S3 = 11	S1 = 01	0	1

$$D1 = \bar{I} \cdot Q0 + I \cdot \overline{Q0} \cdot Q1$$

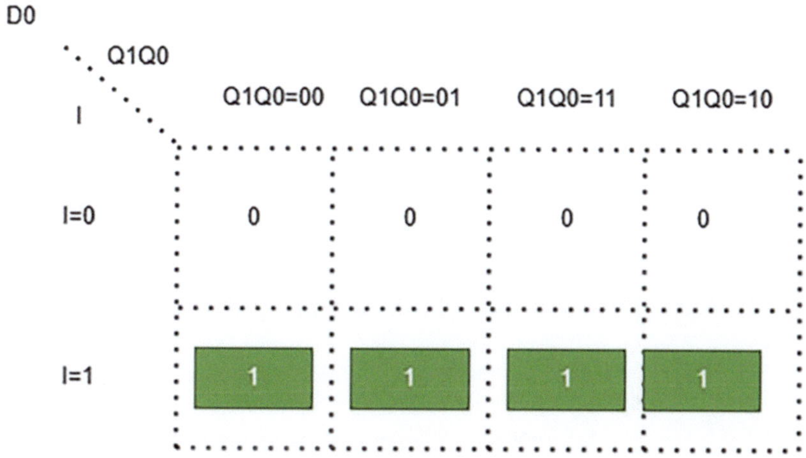

$$D0 = I$$

7. **Output logic**

In the Moore FSM output if function of the present state only (Table 10.10).

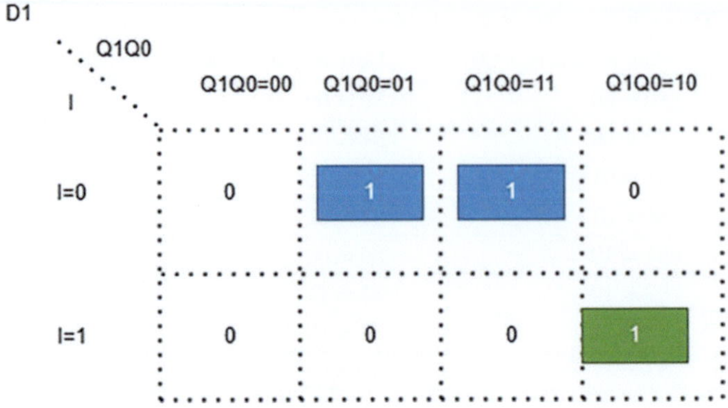

Fig. 10.11 The K-map for the next state logic

Table 10.10 The truth table of the output logic

Input (I)	Present state (Q1 Q0)	Next state ($Q1^{+}$ $Q0^{+}$)	Y
0	S0 = 00	S0 = 00	0
1	S0 = 00	S1 = 01	0
0	S1 = 01	S2 = 10	0
1	S1 = 01	S1 = 01	0
0	S2 = 10	S0 = 00	0
1	S2 = 10	S3 = 11	0
0	S3 = 11	S3 = 10	1
1	S3 = 11	S1 = 01	1

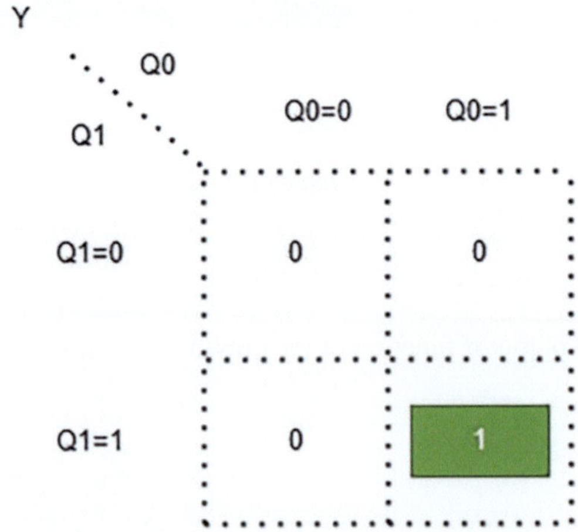

Here $Y = Q1 \cdot Q0$.

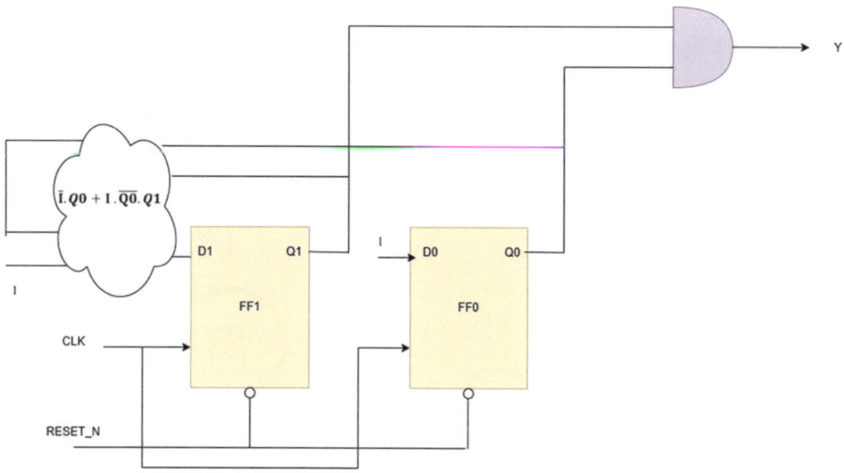

Fig. 10.12 The logic design of the 101 overlapping Moore sequence detector

8. **Let us sketch the FSM Design**

The design is ss shown in Fig. 10.12.

10.6 Mealy FSM Design

Let us consider the design of the sequential circuit to detect the 101 overlapping sequence We can use the following design steps to design the Mealy FSM.

1. **Find the number of states to detect the 101 overlapping Mealy sequence**

Number of states = 3. The states are S0, S1, and S2.

2. **State diagram**

The state transition happens as shown in Fig. 10.13.

3. **Find number of flip-flops**

We will use the binary encoding and the number of flip-flops $= n = \log_2 3 = 2$ We will use the positive edge sensitive D flip-flops.

4. **Reset strategy**

Let us use active low asynchronous reset input reset_*n*. For reset_$n = 0$ the sequence detector holds the default state. For the reset_$n = 1$ the state transition happens on the rising edge of the clock.

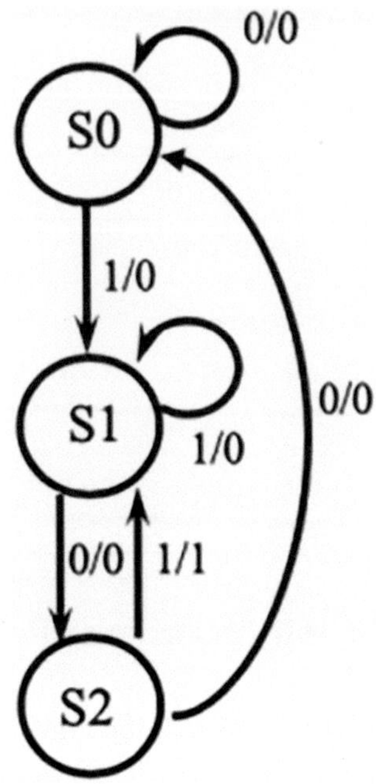

Fig. 10.13 Mealy FSM of 101 overlapping Mealy machine

5. **Let us document the entries in the state table to get state register logic**

The state table of the 101 overlapping Mealy sequence detector is shown in Table 10.11.

So, we need to have two D flip-flop having asynchronous active low reset input reset_n and rising edge sensitive clock.

6. **Use of the excitation table to design the next state logic**

The excitation table consists of the information about the present state, next state, and excitation input (Table 10.12).

Now let us use the K-map to deduce the Boolean equation for the next state logic that is D1, D0 (Fig. 10.14).

Table 10.11 The state table of the 101 overlapping Mealy sequence detector

Input (I)	Present state (Q1 Q0)	Next state ($Q1^+$ $Q0^+$)
0	S0 = 00	S0 = 00
1	S0 = 00	S1 = 01
0	S1 = 01	S2 = 10
1	S1 = 01	S1 = 01
0	S2 = 10	S0 = 00
1	S2 = 10	S0 = 01

Table 10.12 The excitation table of the Mealy 101 nonoverlapping sequence detector

Input (I)	Present state (Q1 Q0)	Next state ($Q1^+ Q0^+$)	D1	D0
0	S0 = 00	S0 = 00	0	0
1	S0 = 00	S1 = 01	0	1
0	S1 = 01	S2 = 10	1	0
1	S1 = 01	S1 = 01	0	1
0	S2 = 10	S0 = 00	0	0
1	S2 = 10	S0 = 01	0	1

$$D1 = \bar{I} \cdot Q0$$

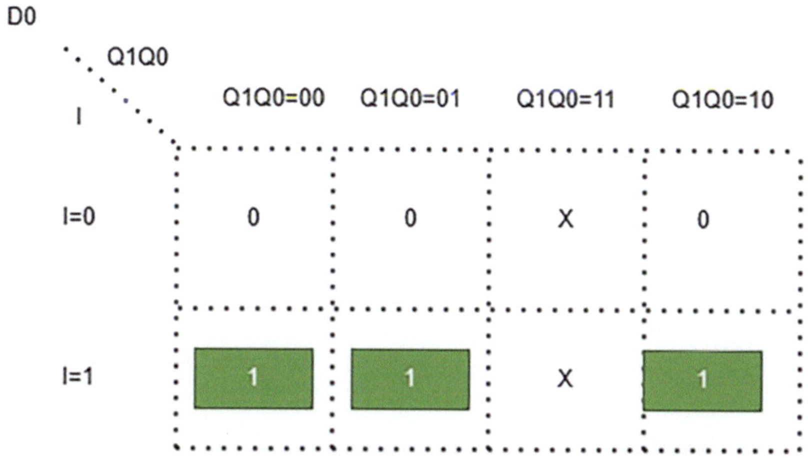

$$D0 = I \cdot \overline{Q0} + I \cdot \overline{Q1}$$

7. **Output logic**

In the Mealy FSM output if function of the present state and input (Table 10.13).

Now let us use the K-map to deduce the Boolean equation for the output logic (Fig. 10.15).

$$Y = i.Q1$$

So, we need to have AND gate as an output logic.

8. **Let us sketch the FSM Design**

The design of the Mealy 101 nonoverlapping sequence detector is shown in Fig. 10.16.

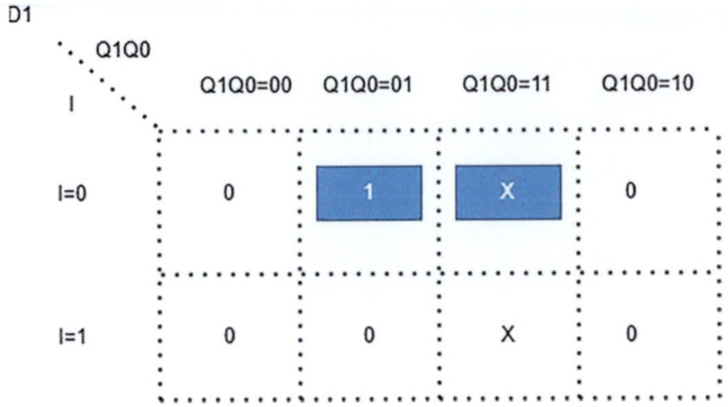

Fig. 10.14 The K-map for the next state logic

Table 10.13 The truth table of the output logic

Input (I)	Present state (Q1 Q0)	Next state ($Q1^{+}$ $Q0^{+}$)	D1
0	S0 = 00	S0 = 00	0
1	S0 = 00	S1 = 01	0
0	S1 = 01	S2 = 10	0
1	S1 = 01	S1 = 01	0
0	S2 = 10	S0 = 00	0
1	S2 = 10	S0 = 01	1

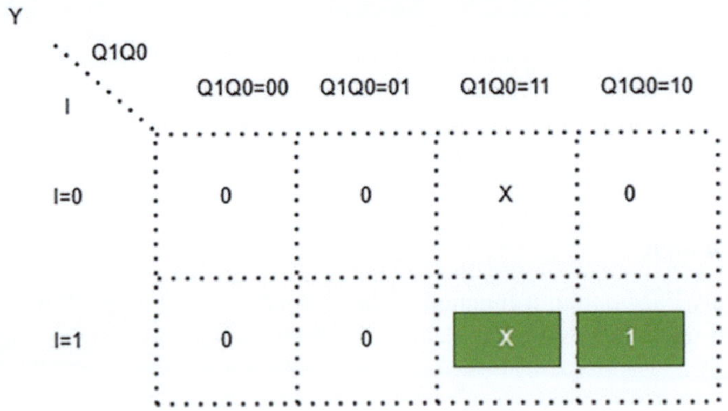

Fig. 10.15 The K-map for the output logic

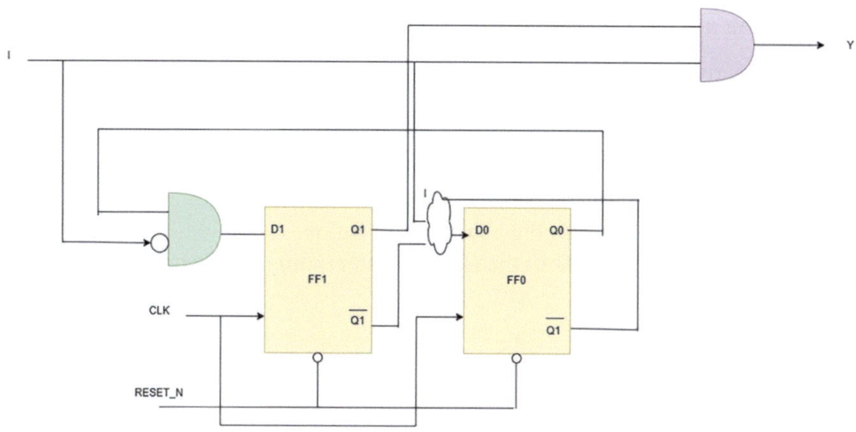

Fig. 10.16 The Mealy 101 overlapping sequence detector

10.7 FSMD

The FSM data path designs are used in most of the SoCs and the objective is to have the separate data path and control path for the design (Fig. 10.17).

As shown in the Figure the FSM controller generates the active high enable output when the sequence 101 is detected and the DATA_OUT=DATA_IN

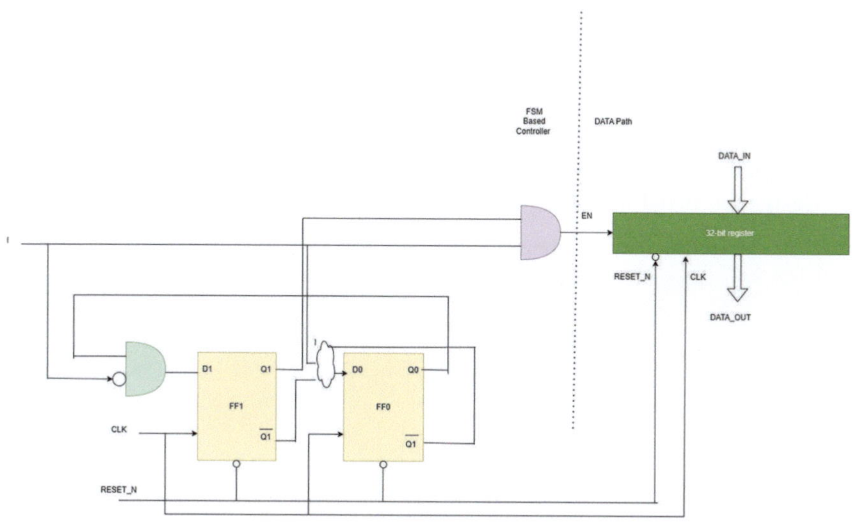

Fig. 10.17 The FSMD

Following are few guidelines to have the FSMD

1. The objective of the logic designer is to design the glitch free FSMs.
2. The FSM-based controllers should have separate data and control path for better area and speed.
3. While designing the FSM to optimize for the area try to use the strategy to eliminate unwanted states.
4. Use the Gray encoding for the power optimization.
5. If the area is not a constraints, then for the better timing use the one-hot encoding.

10.8 Important Highlights

Following are few of the important points to summarize this chapter.

1. In the Moore FSM, an output is function of the present state only.
2. In the Mealy FSM, an output is function of the present state and present inputs.
3. Mealy FSM needs lesser number of states as compared to Moore FSM.
4. The FSM-based design approach is useful to design the sequence detectors and random counters.
5. The FSM can use one of the state encoding method, binary, gray or one-hot.
6. Using the gray encoding the power can be optimized.
7. Using the one-hot encoding method the timing can be improved but it increases the area.
8. The SoC should have the FSM with the data path.

10.9 Puzzle

1. Design the 10101 mealy sequence detector using the gray encoding?
2. Design the 10101 mealy sequence detector using the one-hot encoding?

Chapter 11
Low Power Design and Optimization Techniques

The chapter covers the low power design techniques and practical scenarios. The chapter also covers the power optimization, multiple power domain designs and the design considerations.

11.1 Introduction

The low power design improves the energy consumption and battery life in the modern SoC designs. There are various techniques which are used in the SoC designs to reduce the static and dynamic power and SoC design needs these techniques. The sources of the static power are the DC and leakage current and dynamic power is function of the frequency due to the short circuit current and switching activity. Even following are few of the factors which need to be considered during the design of the low power aware architectures

1. To meet the power supply requirements during the power grid calibration increases the design and manufacturing cost.
2. Packaging and cooling increases the system costs.
3. Shorter battery and device life
4. Shorter design cycle lower device reliability
5. Shorter battery life as we need to have portable plug and play kind of the devices.

As discussed, the power is an important design consideration. Most of we have come across the terms such as switching power, static and dynamic power. The power dissipation should be minimum. Consider design of NOT gate, as shown as the stray capacitance is formed at the output the power is specified as (Fig. 11.1)

$$\mathbf{Power = \alpha CsVDD^2f}$$

V. Taraate, *Advanced Digital Design Techniques*,
https://doi.org/10.1007/978-981-95-4510-0_11

where α = switching factor

Cs = load or stray capacitance

VDD = supply voltage

To have the minimum power the load capacitance should be minimum, the voltage should be minimum but as we can't compromise on the speed of the design we need to balance between the desired power and speed of the design.

There is always the tradeoff between the speed and power and the main goal of the logic design team is to have the balance act to achieve the desired speed and power for the design

Few of the power optimization techniques are listed below

1. Clock gating for the sequential logic
2. Use of low power cells in the design
3. Low power aware architecture design
4. Voltage and frequency scaling

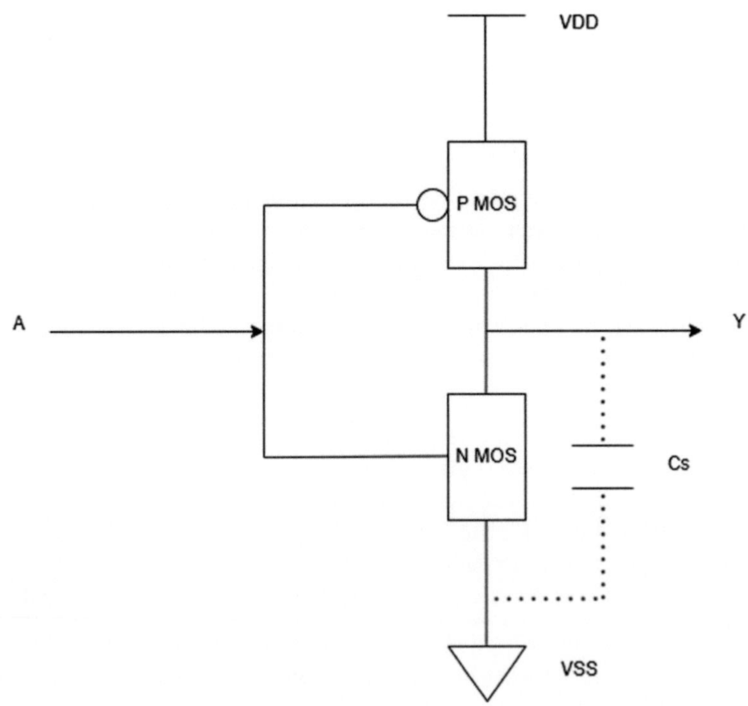

Fig. 11.1 The logic with the stray capacitance

11.2 Low Power Design Challenges

Always there are challenges in the low power aware architecture design and few of them due to switching are documented in this section.

1. The drop in the voltage level and the sign-off is one of the major challenges in the ultra-low power SoC designs.
2. Lower supply voltages have impact on the path timing and they are susceptible to dynamic voltage drop (DVD). Very difficult and challenging tasks to catch the DVD. IR drop analysis is the real challenge.
3. UPF, we use as a standard but still careful design of the power and clock management unit to have the good power is the key challenge. The smooth transition of the power states and supply of the multiple voltage levels to the functional block is one of the important considerations during the design.
4. The logical and electrical switching while passing the clock gating signals is another challenge in the low power design.
5. The clock gating logic and the smoother transition to have the clock enables is another challenge.
6. The power reduction strategies in the test mode versus the normal mode and is one of the challenges.

By considering the above challenges the low power aware architecture should be designed and more time should be invested by the team in the design of the power state machines, power sequencers to have good power and clean timing during the switching. The following section is useful to understand the low power design techniques used in the modern SoCs.

11.3 Clock Gating Cells

The clock gating is one of the important design techniques used in the low power design to reduce the dynamic power. Now let us think how the power is consumed? In the design, power is consumed during the charging of the load capacitance or when the transistor changes its state. The total dynamic power is summation of the transient power and charging power for the capacitance.

Consider the following equation of the dynamic power in the SoC.

$$P_{dynamic} = P_{cap} + P_{transient} = (C_L + C)V_{dd}^2 fN^3$$

In the equation **C** is the chip capacitance, $\mathbf{C_L}$ is the load capacitance, the supply voltage may be for the lower power consumption the Vdd is less than 1.0 V. Most of the modern SoCs has low power aware design architecture and have the power management circuit. The **f** is design frequency, and **N** is the total number of bits switching at the particular instance of the time.

So, by considering the above equation we can have the lower dynamic power by turning off the clock signals to certain functional blocks of the SoC. The reason being the clock gating will not allow switching of the transistors in that particular functional block and reduces the dynamic power.

Consider the clock of 500 MHz and as a designer I know the time duration for which the functional block should be ON or power up in such situation we can turn-on the logic for the specific time duration as required. To achieve this, we can use the enable signal. So, for the enable = 1 and clk active the gated clock is generated and used as clock to the sequential elements that is flip-flops. This reduces the charging and discharging at the clock input of the sequential cell and reduces the dynamic power.

The clock gating logic is designed using the AND gate, the AND of the CLK, CLK_EN and shown in the Fig. 11.2.

If we observe carefully the clock gating by AND gate, we can conclude that, the gated clock is prone to the glitches and practically this is not a viable solution for any kind of SoC. We need to have the dedicated clock gating cells and they are designed by using the latches and AND/MUX (Table 11.1).

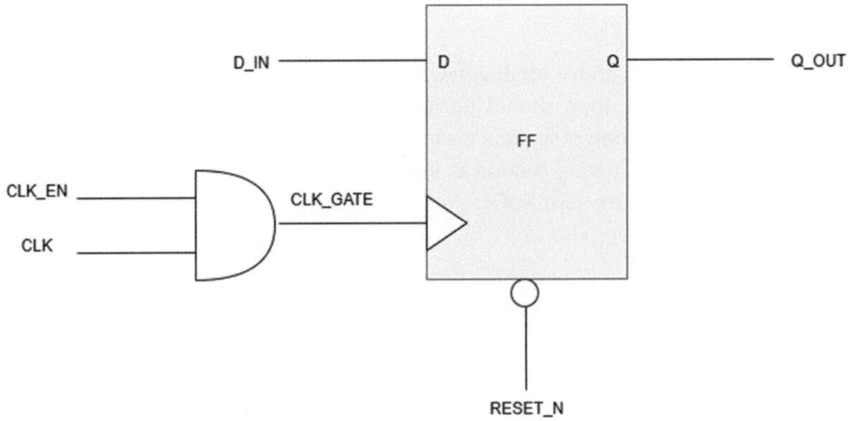

Fig. 11.2 The basic clock gating logic

Table 11.1 The clock gating table

CLK_EN	CLK_GATE
1	CLK ON
0	CLK OFF

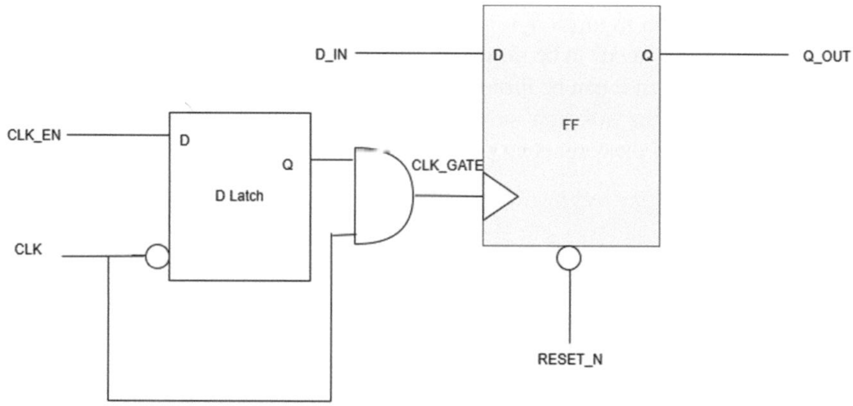

Fig. 11.3 The clock gating cell to have glitch free CLK_GATE

11.4 Glitch Free Clock Gating Cell

As discussed in the previous section the gated clock generated by using only AND gate is prone to the glitches and as it is not viable solution. In the ASIC designs we can use the dedicated clock gating cell to generate the CLK_GATE that is gated clock.

The clock gating cell is designed using the negative level sensitive D latch and AND gate. The AND gate can be replaced by using the MUX to have the clean clock. The clock gating cell to generate the CLK_GATE is shown in the Fig. 11.3.

When the CLK is negative level, the latch is transparent and passes the CLK_EN to the latch output Q. During the positive level of the clock the latch is deactive and holds the previous value. The output of the latch is one of the inputs of AND gate and CLK is another input of AND gate. The CLK_GATE is glitch free and reduces the dynamic power in the design.

11.5 Low Power Architecture Design Considerations

Following are few of the low power design strategies for the reduction of the power.

11.5.1 Power Gating

We can use the power gating to turn-off the power to the non-functional blocks. For the SoC designs if we have the multiple power domains such as PD1, PD2, PD3 … PDn, that is n islands of the power. We need to have the power sequencer and power state machine control to turn-on, turn-off the power of the functional blocks. When the functional block needs to be turned on the power gate should be ON and when it is non-functional then it can be turned-off using the power gating enable control.

We need to have the isolation and retention cells to isolate the power domains and to retain the logic of the respective power-off domain, respectively.

11.5.2 Frequency Gating and Dynamic Frequency Scaling

If we can use the clock gating logic to control the clock to the various functional blocks of the design then we can reduce the dynamic power. Consider the SoC has the processors, memory controllers, bus controllers operating at the different operating frequencies. In such scenarios we can partition the design based on the clock frequencies. The processors in one of the clock domains may be clock domain 1, the memory controllers in second clock domain 2, the bus controllers in the clock domain 3. We can have the clock management unit which can generate the clock gating enable signals as required depending on the functional requirements of the design. This is frequency gating and as the frequency can be scaled down from the higher value this is significantly used to reduce the dynamic power.

11.5.3 Multivoltage (Multi-Vdd) Design

Most of the time we know that the all the SoC functional blocks may not need the same Vdd. If we have Vdd of the 1.8 V and the SoC or chip power is few micro-watts then it can be reduced by using the multivoltage design. If we can create the voltage islands of 1.8, 1.2, 0.9 and 0.6 V then we can get significant improvement of the power. As power is directly proportional to the square of the voltage the chip power can be optimized using the multiple voltage islands.

11.5.4 Dynamic Voltage and Frequency Scaling

The multiple voltage islands will generate the fixed Vdd and using the software or firmware control we will not be able to change the fixed voltage of that specific island. In such scenario there will never be the further power optimization. To have the power optimization we can dynamically change the Vdd and dynamically change the frequency as required. In the low power state depending on the power needs the Vdd can be further reduced using the firmware or software control and even frequency can be reduced using the firmware or software control. This type of technique is heavily deployed in most of the SoCs and called as the dynamic voltage and frequency scaling.

11.6 The UPF

Each complex million gate SoC uses the Intellectual Property (IP) and to have the lower power consumption we need to transit between the power modes. That is power-off, power-on, sleep mode. Each IP can have multiple power domains and can transit between the power-off, power-on modes to reduce the power. We may need to restore the values of the memory cells then we need to have the retention control and retention cells. To avoid the leakage and data corruption we need to deploy the isolation cells from the power-off domain to the other domains. The SoC should have the dedicated power management control unit and we use the power specification formats such as CPF/UPF. The power specification formats are used to design the power management circuits with the objective to have the lower power consumption.

CPF is Common Power Format
UPF is Unified Power Format

What we need is

1. ***Level shifters***: For the different voltage domains or blocks which are operating at different voltage levels we need to have the level shifters.
2. ***Power gates***: Power domains or blocks with a separate power supply that can be turned on or turned-off
3. ***Supply nets***: Multiple supply nets and connections having the names according to the requirement of the standard.
4. ***Power switches***: The large switching transistors to force the logic in the power-off state.
5. ***Isolation Cells***: The isolation cells at the output of the power domains which is on.
6. ***Retention cells***: To retain the flip-flop outputs in the power domain which is in the power-off state.
7. ***Always on Logic:*** The always on logic cells and always on path logic which is power-on state when the supply is turned off.

So, we need to have the power management unit to have the good power and smooth transition of the power modes. The following section is useful to understand the power management unit and design of the efficient power management unit to have smoother transition from power-on to power-off modes.

11.7 The Power Management Unit

The power management unit design is responsibility of the SoC architect as the goal is to have low power designs. The architecture of the power management unit uses various functional blocks and this section is useful to understand the details of the power management unit.

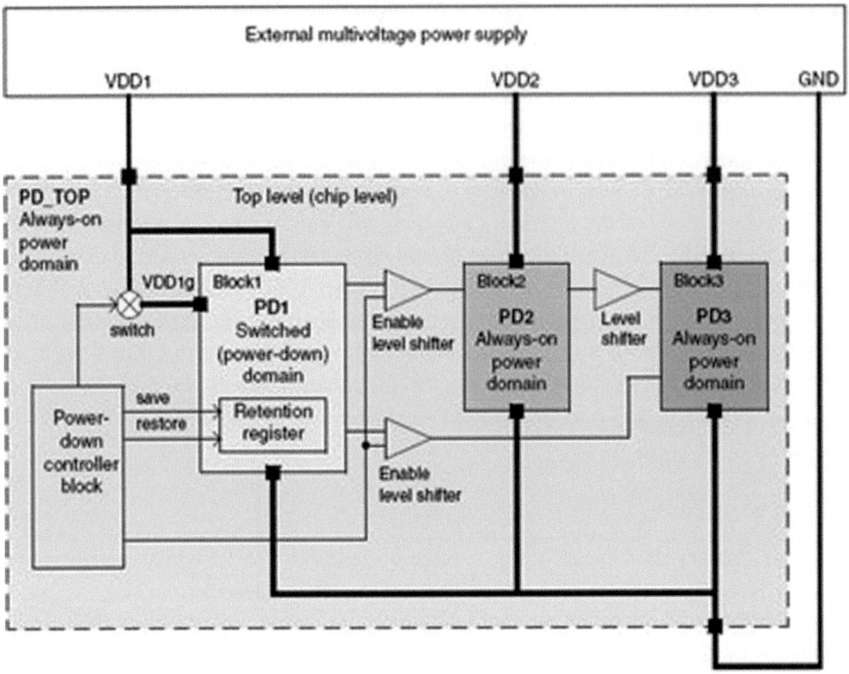

Fig. 11.4 The power management unit

As shown in the Fig. 11.4**, the power management unit is designed using the following**

Power Domains: The design has the multiple power domains Pd1, Pd2, Pd3. Using the power switches the blocks can be turned on or off.

Multivoltage power supply: For the different power domains the supply voltage of the VDD1, VDD2, VDD3 is applied depending on the required power calculations. The level shifters are needs between the power domains and deployed and controlled by the power down controller block. The power down controller block is the power FSM and takes care of the smoother transitions for the various power modes.

Power Switches: The power switches are used to enable or disable the power gates in the power management unit.

Retention Registers: The retention cells or retention registers are used in the power down domain PD1 to retain the state of the memory or flip-flops during the power down mode.

Isolation Cells: To isolate the power-off domain from the power-on domain the isolation cells are placed at the boundary of the power-off and power-on domain.

11.8 Low Power Aware SoC Designs

For the SOC Design we need to have the following

1. ***PLL and Clock Circuits***: The Phase Locked Loop(PLL) in thc lock range generates the good clock and even the pll_locked signal from the PLL can be used as good clock signal to the design. Depending on the design requirements the PLL should generate the high frequency clocks and these clocks should be used as an inputs by the clock and reset management block to generate the slower clocks depending on the design requirements.
2. ***The clock and reset management unit***: All the clocks need to be captured and the faster and slower clocks need to be generated by the clock management unit. The PLL generates the main high frequency clock and the internally generated clocks need to be designed using the configurable clock dividers to have clean clock paths. The clock gating strategies need to be deployed for the reduction of the dynamic power. The active low reset assertion on the asynchronous reset input and need to deploy the reset synchronization for the deassertion. For more details of the reset synchronizer refer Chap. 9.
3. ***Power Management Unit***: The SoCs have the power management unit which consists of the

    i. Power sequencer and state machines
    ii. The power gate enables
    iii. The power islands
    iv. The level shifters
    v. The isolation cells and isolation controls
    vi. The retention cells and retention control

The power management unit is responsible for the power mode transitions and power control generations for the smooth transition of the power states as required. The main objective is to have lesser dynamic power dissipation and the low power aware SoC design. Use the UPF or CPF during the low power aware design and simulation.

4. ***Functional Blocks***: Depending on the design requirements the SoC can have one or more than one processors, memory controllers, video encoders and decoders.
5. ***Multiple clock domains and CDC***: The design has the multiple clocks and the required CDC techniques and synchronizers need to be deployed to have the data convergence and to avoid the metastability while exchanging the data or control signals from one of the clock domain to the another clock domain.
6. ***Internal Memory***: The SoC should have the config and status registers and scratch pad memory as required. The SoC can have the dedicated control and status registers and should be mapped to the functional blocks as required depending on the architecture design strategies.
7. ***Bus interfaces***: The dedicated serial and parallel bus interfaces such as UART, I3C, SPI, PCIe Gen 5/6, APB, AXI, AHB as required should be used.

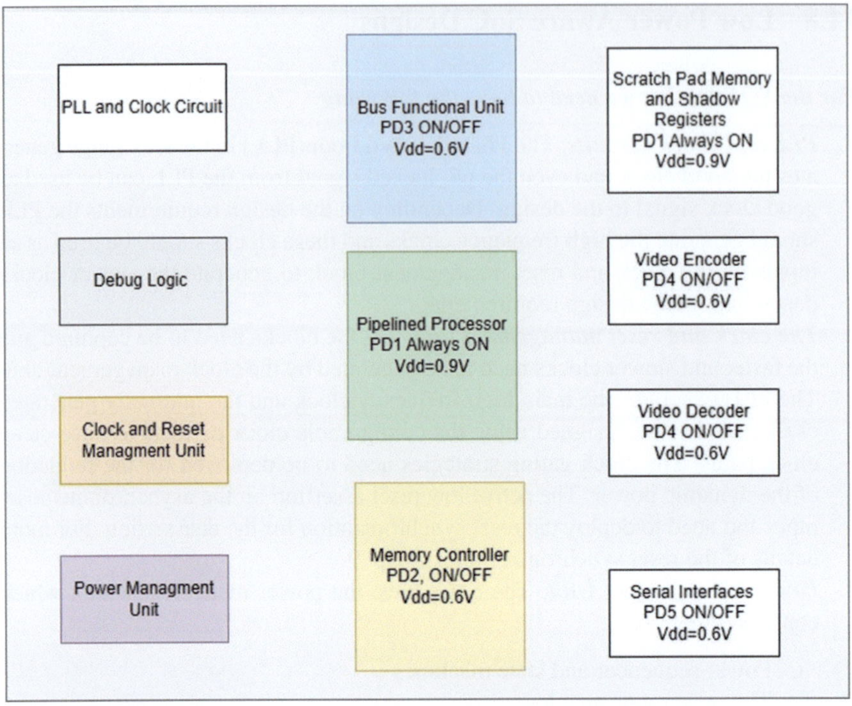

Fig. 11.5 The SoC TOP

8. ***Debug Logic***: Most of the SoCs have the debug logic and the debug control registers. During the debug mode the design data path functionality can be checked and objective is to confirm the functional correctness of the data paths. The debug mux, demux chains need to be deployed according to the requirements and even the JTAG based testing should be used to test the functional correctness of the data path and the data integrity checks.

The SoC-Top is shown in the Fig. 11.5.

11.9 Important Highlights

Following are few of the important points to summarize this chapter.

1. The power management unit to have smoother transition of the power modes should be used in the ultra-low power SoCs
2. The clock gating is used to optimize the dynamic power
3. The CPF and UPF are two power formats used to have the low power aware design and verification

4. The dynamic voltage and frequency scaling is one of the important techniques used to reduce the power.
5. The power management unit has the power controllers, power switches, level shifters, isolation and retention cells
6. The retention state registers are used to retain the value of the flip-flops during power-off state.
7. The isolation cells are used to isolate the power-off domain from the power-on domain.

Chapter 12
DFT-Friendly Designs and Practical Scenarios

The chapter covers the application of the Chaps. 1–12 in the digital system design. Chapter is useful during the design of the digital system which has IPs, SoC components. Also, the main focus of this chapter is to have detail understanding of the DFT-friendly designs and test circuits used in the design and their optimization to have higher performance of the digital system.

12.1 Scan Flip-Flop

The design should be testable and we should have the DFT-friendly design which is controllable and observable. During the design stages we always keep in mind that the design should have controllability and observability. That is inputs should be controllable and outputs should be observable to test the circuits.

The DFT is design for the testability and all the modern SoC designs has the test and debug control. We can have the partial scan chains and full scan chains using mux-based scan flip-flops.

The scan flip-flop uses the scan input in the test or scan mode and passes through the D input of sequential elements to generate the scan output. The mux-based scan cell is shown in the Fig. 12.1.

As shown in the Fig. 12.1, the mux and flip-flop are used to have the mux-based scan cell.

As described in the Table 12.1, for the SCAN_EN = 1 the output of the scan flip-flop is equal to SCAN_IN and during the normal mode operation the output of the scan flip-flop is equal to the DATA_IN.

If we have the clock path 1 and has the 10000 sequential elements then we need to have the scan flip-flops to form the scan chain of 10,000 flip-flops. This mechanism is used to debug or test the design and we can complete connectivity checks as the design is completely controllable and observable.

V. Taraate, *Advanced Digital Design Techniques*,
https://doi.org/10.1007/978-981-95-4510-0_12

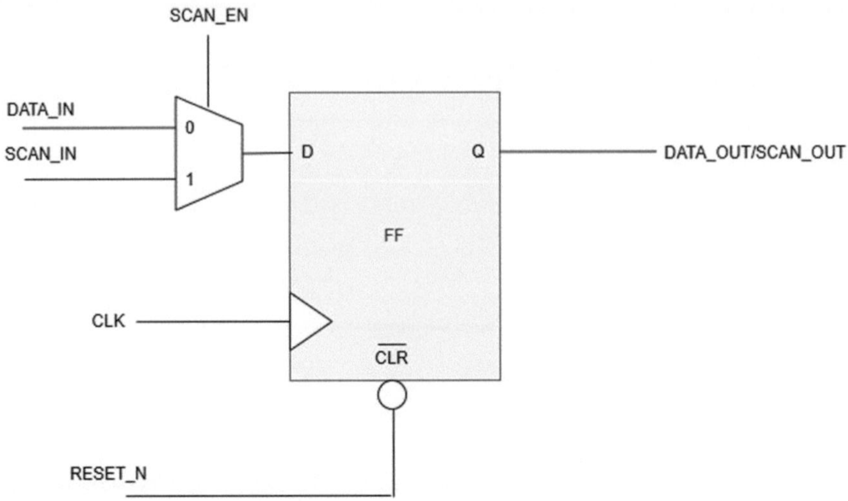

Fig. 12.1 The mux-based scan cell

Table 12.1 The truth-table of the scan flip-flop

SCAN_EN	DATA_OUT/SCAN_OUT
1	SCAN_IN
0	DATA_IN

For the n stages the mux-based scan chain is discussed in the following section.

12.2 Scan Chains in the Design

In the modern SoC designs we have the multiple clock groups and design is partitioned with the goal to check for the functional correctness and to test it during the test mode. Effectively the manufacturing faults and fault coverage needs to be checked before the production of the chip. By considering this requirement we use the DFT techniques and we improve the controllability and observability of the design.

The goal of the architect and logic designer is to have the DFT-friendly designs and architecture as it will be used during the RTL design phase.

Consider we have the design of multimillion logic gates. What we need to do is

1. Have the logical partition of the design based on the clock and power requirements.
2. Have the clock groups based on the clock domains.
3. Have the strategy for the testability. ‘
4. Insertion of the scan chains and check for the fault coverage.

Table 12.2 The scan chain

SCAN_EN	DATA_OUT/SCAN_OUT
1	SCAN_IN
0	Operation on the DATA_IN

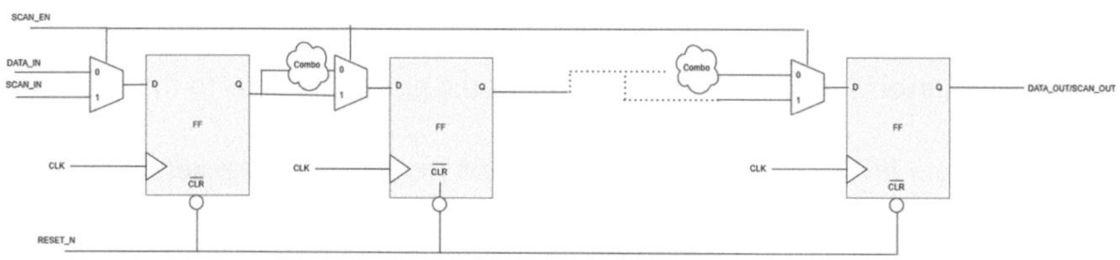

Fig. 12.2 The mux-based scan chain

5. The complete testable design in the debug mode with objectives to find the faults in the designs.

By considering this, if we have few lac sequential elements in the clock group 1 then we can use the scan-flip-flops to have the scan chain. Table 12.2 describes the relationship between the inputs and outputs.

Normal Mode: During normal mode the SCAN_EN = 0 and the data_in passes through the reg-to-reg path to generate the output depending on the functionality of the design. If we consider the reg-to-reg path then output of flip-flop is used by combo logic to generate the input of the next stage flip-flop.

Test Mode: During the scan or test mode the SCAN_EN = 1 and the design uses the SCAN_IN input and forms the shift register as the combo logic is bypassed. The mux output is SCAN_IN and travels through the n stage scan chain. This improves the controllability and observability in the design and we can check the connectivity and report the faults with the fault coverage.

The mux-based scan chain is shown in the Fig. 12.2.

As we know that the design should be DFT friendly and should be low power design. By considering this scenario we need to think about the

1. Power aware scan chains.
2. Test synthesis for low power aware designs.
3. Test insertions for the low power designs.
4. Test strategies for the low power designs.

The subsequent section is useful to understand these techniques.

12.3 Scan Chains for the Low Power Designs

It is basic question which I need to answer to understand why we need to have the power aware scan chains? What are the issues and what should be our objectives?

1. In the multiple power domains, we have the always On and ON–OFF power domains and we know that, during the test mode it is not possible for the test clocks, control signals to travel from one of the power domains to the power off domain.
2. We have the level shifters, isolation cells and the retention registers depending on the power domain architecture.
3. So, the basic mechanism is having the partitioning of the design based on the power domains and then create the power aware scan chains.
4. Each scan chain can have the dedicated scan I/Os, test clocks and control signals.
5. This will have higher IOs and test resources and expensive but as each power domain has the dedicated power aware scan chain the testability will be easy.

Consider we have three power domains PD1, PD2, PD3 as shown in the Fig. 12.3. Here care is taken to utilize the scan resources completely to have the balanced scan chain in the test mode.

The power aware scan chain implementation for the three power domains PD1, PD2 and PD3 is shown above. As shown for the multiple Vdd power domains the bypass multiplexers are used.

Due to the use of the bypass multiplexers, if the power domain PD1, PD3 are ON then the bypass multiplexer bypasses the PD2 as it is in the power down mode.

Following are few of the guidelines for the power aware DFT

1. Each scan chain in the design should be made power aware.
2. There should be the low power and test considerations in the design.

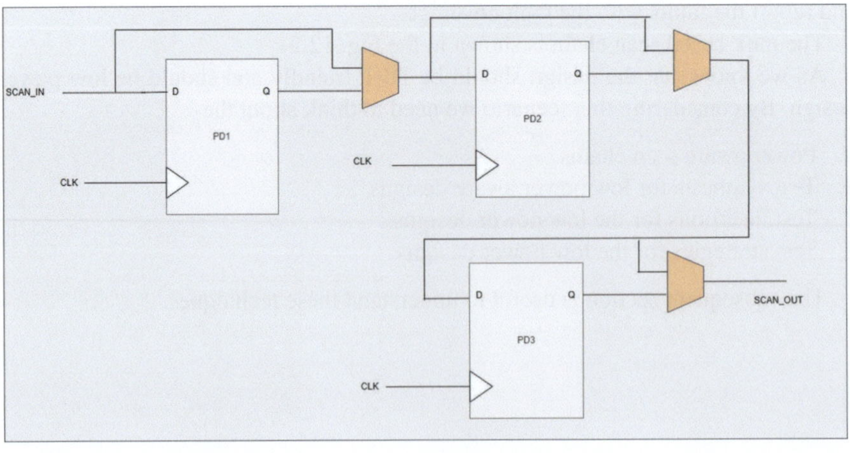

Fig. 12.3 Power aware scan chain

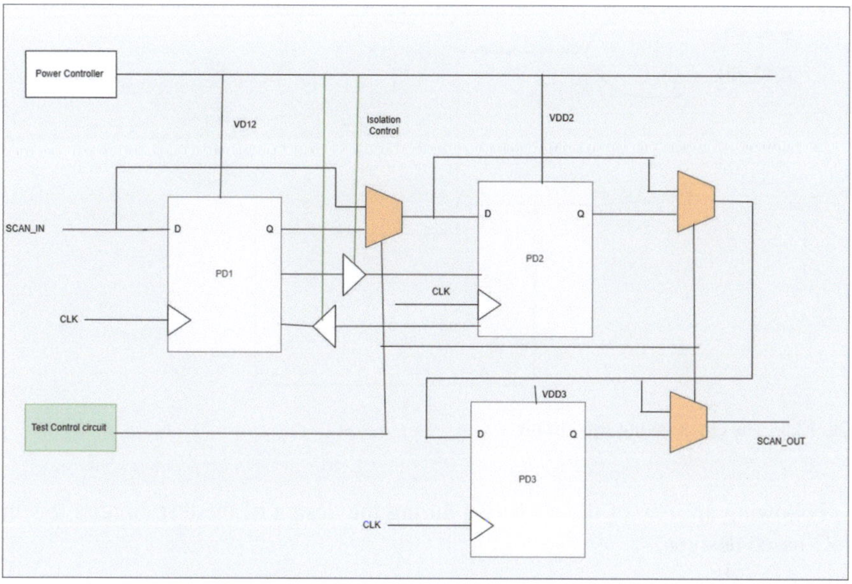

Fig. 12.4 The test controller for the low power aware scan chain

3. The efficient use of the DFT friendly and low power aware architectures need to be deployed.
4. Check the toggling activity and decide the power budget.
5. Have the strategy for the clock and reset controllability.
6. Use the DFT insertion tools and improve the testability of the design by using the Power Test Access Mechanism (Fig. 12.4).

1. We need to have the power controller and test control circuits to have the low power aware scan chain.
2. The test controller controls the scan chain bypass depending on the power ON–OFF sequence.
3. The isolations cells are used to isolate the PD1 from the PD2.

12.4 The Power Management Circuits and Test Scenarios

The main goal and objective of the architect is to have design with the improved fault coverage and it should be easy to test.

We can have the test logic to test the

1. Clock gating cells.
2. Testing of the power switching.
3. Testing of the combo logic in the test mode to have improved observability.

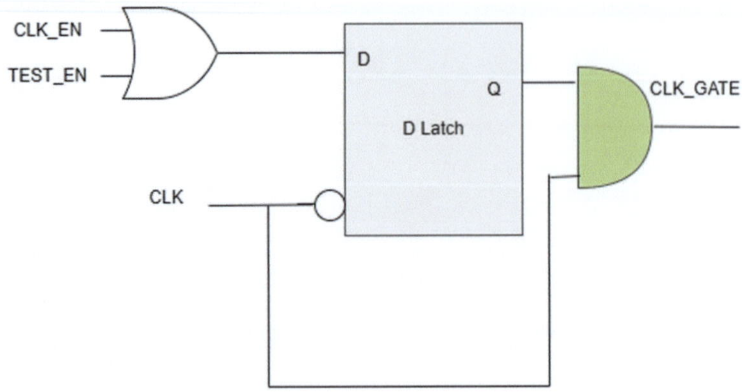

Fig. 12.5 The clock gating test circuit

Following are few of the scenarios during the design of the test circuits for the SoC-based designs.

12.4.1 The Clock Gating Test

The clock gating cell uses the latch to avoid the glitches. The clock gating cells are used to minimize the area and to save the dynamic power. During the normal mode of operation, the design generates the gated clock.

During the test mode the TEST_EN = 1 and used to enable the clock gating logic and we can perform the clock gating test.

The TEST_EN is controllable and the CLK_GATE output from the clock gating cell is observable. This feature improves the controllability and observability of the design (Fig. 12.5).

How to improve the observability in the design?

We can use the observable outputs to improve the observability in the design as shown in the Fig. 12.6.

12.4.2 Power Switch and Testing

To reduce leakage power dissipation the power switches are commonly used in SoC designs.

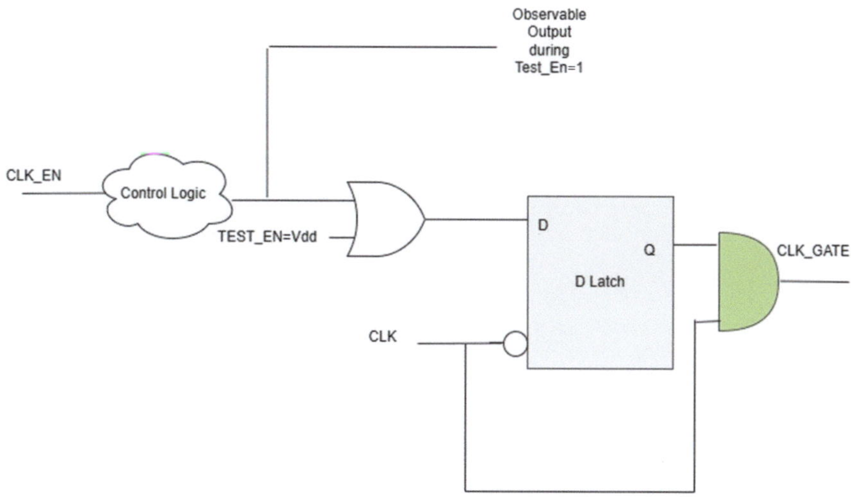

Fig. 12.6 Observability during the test mode

12.4.3 Power Switch and Testing

To reduce leakage power dissipation the power switches are commonly used in SoC designs. We need to have the power switches to enable the power gating (Fig. 12.7).

Using the power switch the functional blocks can be turned on or off as required. We can have the header and footer power switches and are controlled by the active low or active high-power gate enable signal.

For more information on the test scenarios of the power switch refer the DFT techniques and test circuits.

12.5 Important Highlights

Following are few of the important points to summarize this chapter.

1. The test circuits are used to improve the controllability and observability in the design.
2. The clock gating cells are used to reduce the dynamic power.
3. The power switches are used to reduce the leakage power.

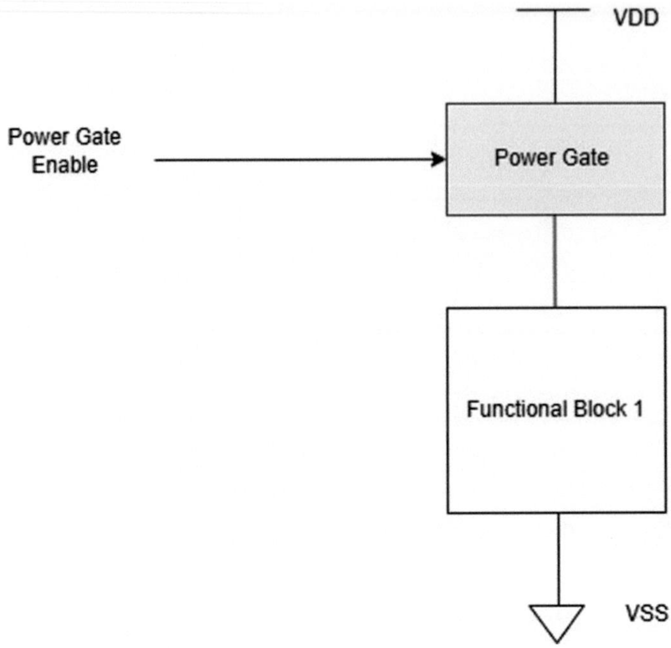

Fig. 12.7 The power gating

Chapter 13
System Design Case Study and Guidelines

The chapter covers the application of Chaps. 1–12 in the system design. Chapter is useful during the design of the digital system which has processors, IPs, SoC components. The design guidelines are discussed in this chapter.

13.1 Pipelined Processor Micro-architecture

The low power aware pipelined controller design is discussed in this chapter. If we consider the processor without pipelining then it performs single operation at a time. Hence, the cycle consumption and power is very high for the design.

Now let us consider the 32-bit pipelined processor logic. The processor has the pipelined stages:

1. Fetch
2. Decode
3. Execute
4. Store.

If we consider the following instructions (Table 13.1).

Let us consider these 4 instructions in the instruction RAM, if every instruction needs 4 cycles to perform the operation, then we need to have 16 clock cycles. The overall cycle consumption is 16 clock cycles.

If we consider the four stage pipelined processor then the processor performs the operation in pipelined manner and needs only 7 clock cycles as shown below (Table 13.2).

Figure 13.1 describes the pipelined logic of the four stage pipelined processor.

Following are the few important steps to be followed during the design

IO Interfaces

V. Taraate, *Advanced Digital Design Techniques*,
https://doi.org/10.1007/978-981-95-4510-0_13

Table 13.1 The instruction set of the 32-bit processor

Opcode S2 S1 S0	Operation
000	A
001	A+B
010	A−B−1
011	A−1
100	A OR B
101	A XOR B
110	A AND B
111	NOT A

Table 13.2 The pipelined instruction execution

Instruction	Fetch	Decode	Execute	Store
PIPO	PIPO	X	X	X
ADD	ADD	PIPO	X	X
SUB	SUB	ADD	PIPO	X
DEC	DEC	SUB	ADD	PIPO

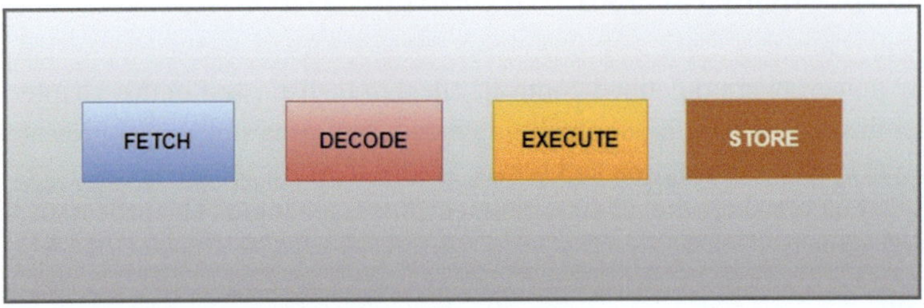

Fig. 13.1 The four stage pipelined processor stages

1. Have the top-level IO signals and description of them with the IO size and details.
2. Here data width is DATA_SIZE = 32bits and ADDR_SIZE is depending on the opcode. For the 8 instruction RISC processor we can have the 3-bit opcode and for the 64 instruction RISC processor we can have the 6-bit opcode (Table 13.3).

The top-level interfaces are shown in Fig. 13.2.

Clock and Reset

The following table gives information about the top-level IO signals with the single clock and active low asynchronous reset (Table 13.4).

The micro-architecture of the processor

The Fig. 13.3 is the micro-architecture of the processor

The design has the following functional blocks

Table 13.3 IOs of the pipelined processor

Sr. No.	Signal Name	I/O	Width	PIN Description
1	Operand 1	I	DATA_SIZE	First Operand Input data
2	Operand 2	I	DATA_SIZE	Second Operand Input data
3	Opcode	I	ADDR_SIZE	Opcode of Instruction
4	Clock	I	1	Clock input
5	Reset	I	1	Reset input
6	Operation output	0	DATA_SIZE	Output data after execution
7	Flag output	0	8	Flag output to indicate overflow etc.
8	Address output for External memory	0	ADDR_SIZE	Address line to address external memory ("It is currently not included in RTL design")

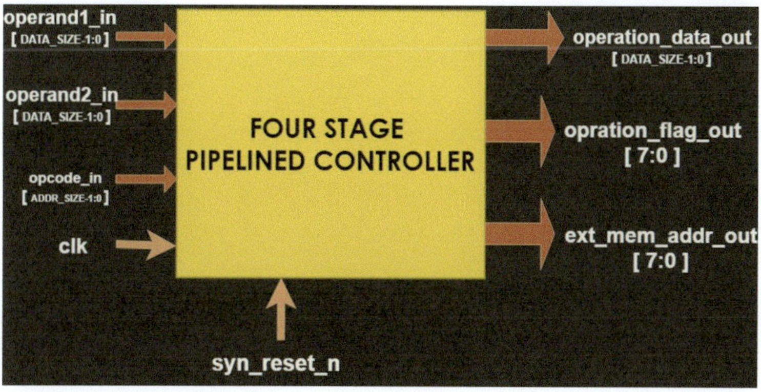

Fig. 13.2 The top-level IO interfaces

Table 13.4 The clock, reset. and IOs of the processor

Sr. No.	Signal Name	I/O	Width	PIN Description
1	Operand 1	I	DATA_SIZE	First Operand Input data
2	Operand 2	I	DATA_SIZE	Second Operand Input data
3	Opcode	I	ADDR_SIZE	Opcode of Instruction
4	Clock	I	1	Clock input
5	Reset	I	1	Reset input
6	Operation output	0	DATA_SIZE	Output data after execution
7	Flag output	0	8	Flag output to indicate overflow etc.
8	Address output for External memory	0	ADDR_SIZE	Address line to address external memory ("It is currently not included in RTL design")

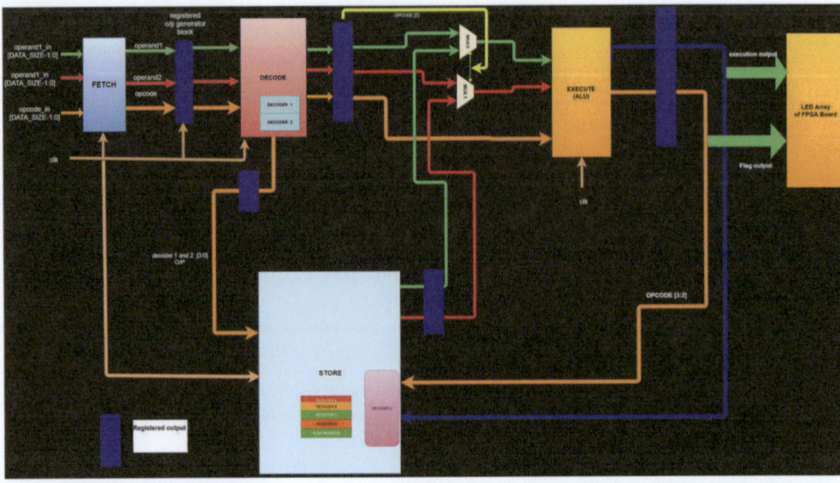

Fig. 13.3 The pipelined processor micro-architecture

1. Fetch block
2. Decode block
3. Execution block
4. Store block
5. Decode2 inside store block
6. Mux base selection logic for operand selection.

Design the sub-blocks using the techniques described in Chaps. 1–12 to have the efficient processor. Use the following guidelines.

13.2 SDRAM Controller Micro-architecture Design

Microprocessor performance has improved at a rate of 60% per year, while the DRAM latency and bandwidth improve slowly.

Memory Controller is the interface between host and physical RAM who works according to instruction given by the host.

It performs specific operation on the physical RAM with the use of control signals.

This interface generally consists of an address input, a data input, a data output, and some control signals to specify a read/write.

Synchronization

Receiving all kind of requests and scheduling the memory access according to the priority of the request.

In order to solve the problem of address conflicts.

To facilitate and implement multi bank concurrent memory access request which are divided into four request group.

Increase in the use of the memory requires smart communicator.

Specifications

Consider the following specifications of the SDRAM

Size: 64 MB

Clock frequency: 100 MHZ (Clock period = 10 ns).

Host Address bus width: 26.

Column address width: 10 (2^10 = 1024 column).

Row address width: 14(2^14 = 16,384 rows).

Bank address width: 2(2^2 = 4 banks).

SDRAM data bus size: 8

Burst length: 2

Cas latency: 3

Refresh time: 64 ms.

No of input and output pins: 21.

IO Interfaces of the SDRAM Controller

The IO interfaces of the SDRAM controller are shown in Fig. 13.4.

The SDRAM controller supports the following commands (Table 13.5).

Micro-architecture

The micro-architecture of the SDRAM controller is shown below and has the following functional blocks.

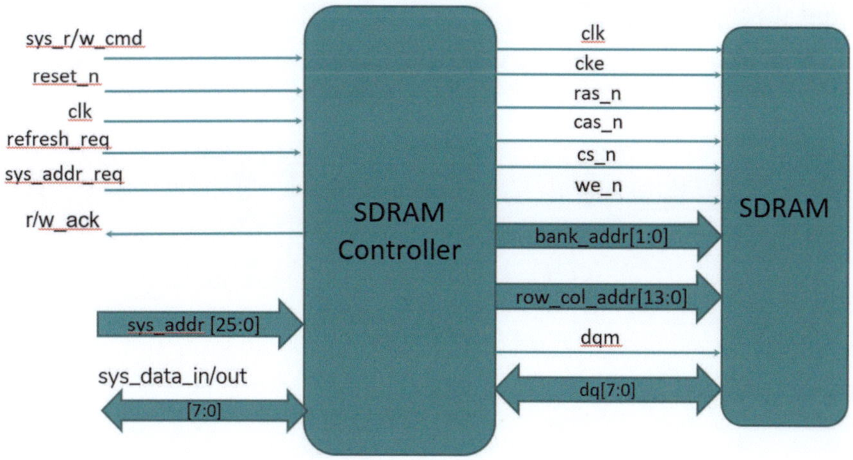

Fig. 13.4 The IO interfaces of the SDRAM controller

Table 13.5 The command table of the SDRAM controller

Command	Sign	cs_n	we_n	cas_n	ras_n
No operation	NOP	L	H	H	H
Load mode reg	LMR	L	L	L	L
Precharge	PRE	L	L	H	L
Read	READ	L	H	L	H
Write	WRITE	L	H	L	L
Active	ACT	L	L	H	H
Auto Refresh	REF	L	L	L	H

1. Counting and timer logic
2. The command buffers
3. The signal generator logic
4. Initialization FSM
5. Command FSM
6. Address decoding logic to select the respective bank (Fig. 13.5).

To design the functional sub-block, use the techniques discussed in Chaps. 1–12.

Following few sections are useful to understand the low power design considerations and DFT friendly designs.

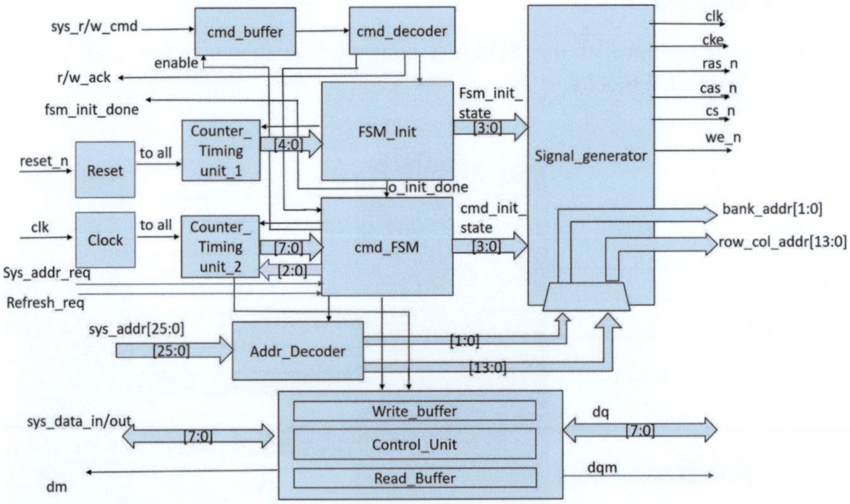

Fig. 13.5 The micro-architecture of the SDRAM controller

13.3 Low Power Design Considerations

Use the following design recommendations to optimize the power.

1. Use the gated clocks in the design to minimize the dynamic power.
2. Use the low power aware design techniques that is dynamic voltage and frequency scaling to optimize the design power.
3. Use the lower technology nodes for the low power high-speed designs.
4. As there is tradeoff between the power and speed use the UPF or CPF during the design.
5. Have the dedicated power management units in the design.
6. Have the separate clock and reset management unit
7. Have the synchronizers while passing data from the clock domain I to clock domain II.
8. Use low power standby modes to reduce the power consumption.
9. Use the parallelism in the design and use the pipelined architecture to have improved power performance.
10. Keep the voltage levels as low as possible as it reduces the overall power consumption.
11. Use the instruction and data caches in the design.

13.4 DFT-Friendly Design Considerations

The design should be completely controllable and observable and we can deploy following strategies to have the DFT friendly architecture and design.

1. ***Clock gating***: If the clock is gated then it should be override by the test logic (Fig. 13.6).

The above technique is prone to the glitches and use the dedicated clock gating cells in the design. Here For the TEST_MODE = 1 the enable is override by the test logic and generates the test clock for the design.

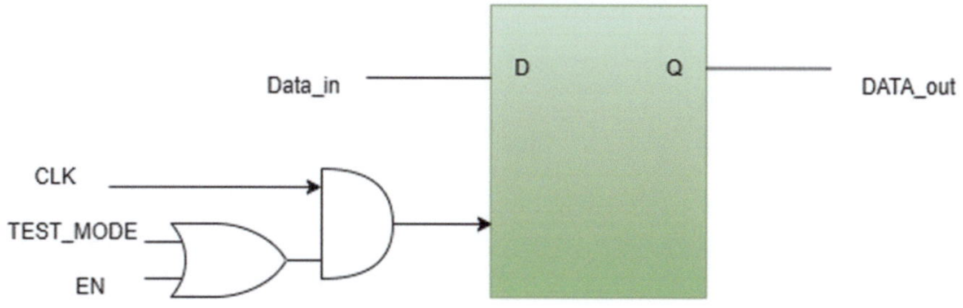

Fig. 13.6 The gated clock

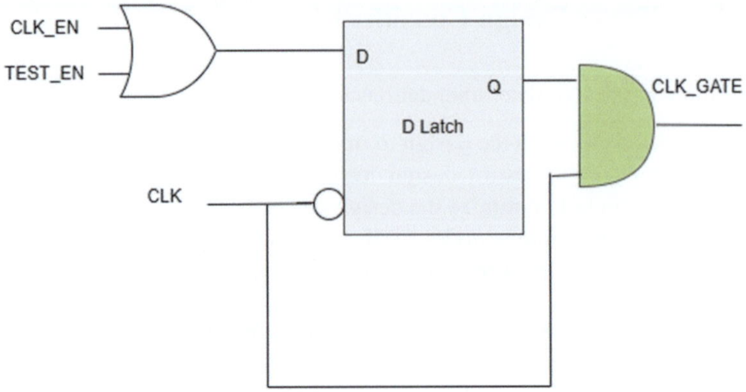

Fig. 13.7 The DFT friendly clock gating logic

As discussed in the previous chapter use the following logic with the test features (Fig. 13.7).

2. ***Bypass the internally Generated clock***: The issue is the controllability of the generated clocks and hence during the DFT friendly design bypass the internally generated clocks using the test mux logic.

As shown in the above figure the design is not DFT friendly and we can't bypass the internally generated clock. We can have the following bypass logic to have DFT friendly design (Fig. 13.8).

3. ***Clock and data path:*** If the clock is used to enable the data path logic for some design specific requirement, then it can't be controllable. Consider following design (Fig. 13.9).

We need to have the testability feature with the improved controllability and observability (Fig. 13.10).

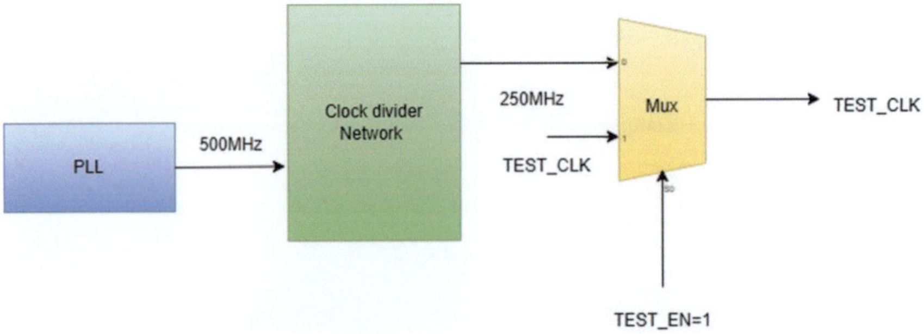

Fig. 13.8 The bypass strategy during test mode

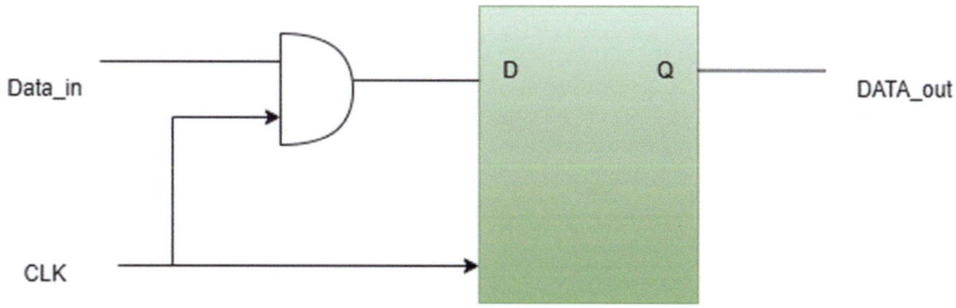

Fig. 13.9 The clock to control data without test logic

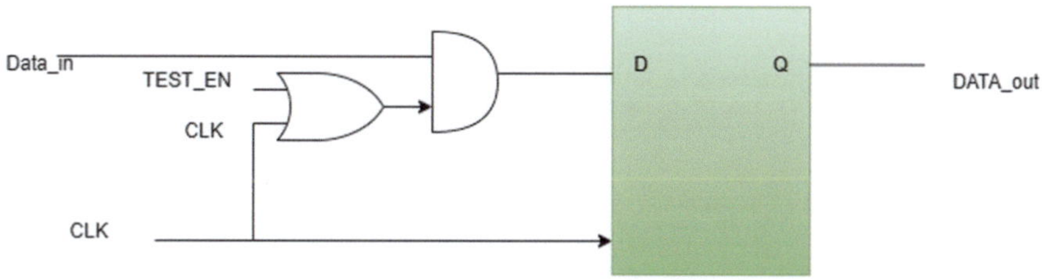

Fig. 13.10 The clock to control data with the test logic

As shown in the figure the design uses the TEST_EN during the test mode and we can control the output of the design.

4. ***Deploy the reset test logic***: In the DFT friendly designs the clock and resets should be completely controllable and to achieve this we can have the RESET mux as shown in Fig. 13.11.

As shown in the Fig. 13.12: The reset test logic, during the test mode the TEST_EN = 1 and the TEST_RESET is used to clear the register.

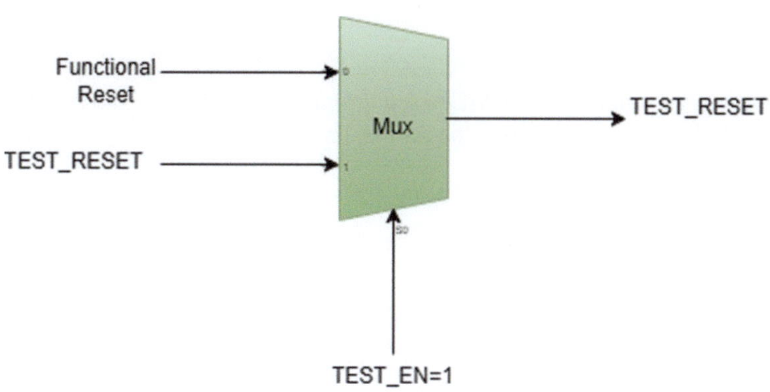

Fig. 13.11 Test mux

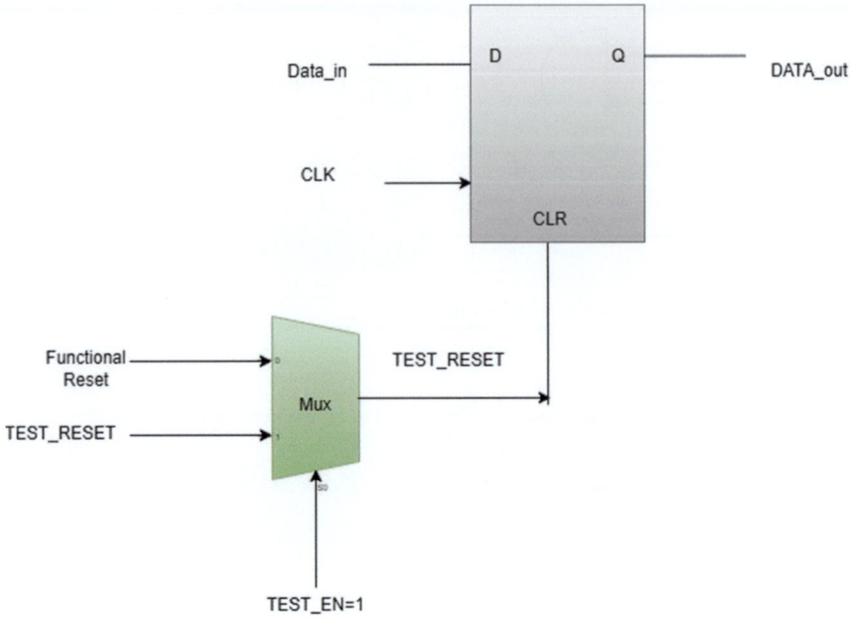

Fig. 13.12 The reset test logic

5. ***Mixed signal designs***: If the design has the analog and digital blocks then we need to consider the following
   (a) The digital logic and interfaces should be fully controllable and observable.
   (b) Use the collar to safe guard the analog blocks.
   (c) Have a strategy of the low power design to keep the low power mode enabled during the testing phase.

13.5 AI ML-Based Design

Now as we need to have the AI/ML based designs we can choose the AI/ML based FPGAs. Modern FPGAs supports the AI/ML based features and they are used in the

(a) Industrial applications
(b) Automotive'
(c) SoC designs
(d) IOT
(e) Signal and speech processing
(f) AI accelerations.

The need of the AI/ML based FPGAs As we know that the complexity of the algorithm has increased so much and high computational algorithm demands the high speed and the low power. By considering this we need to use the modern FPGAs with the AI/ML based application and support.

Advantages of the FPGA-based AI computation

High Performance: If we compare the FPGA performance with the traditional CPUs and GPUs then multiple FPGAs can perform the AI/ML based application to achieve the desired speed.

Power/Energy Efficiency: For any kind of the complex algorithms the FPGA AI accelerators can improve the power and energy efficiency as the AI/ML based processing engines are part of the FPGA.

Flexibility and Adaptability: Because FPGAs are reprogrammable, designers can tailor the hardware to a particular AI application. FPGAs are perfect for quickly growing AI applications because of their versatility, which enables quick upgrades and alterations.

Low Latency: The FPGA based AI accelerators have the lower latency and performs the high-speed computation.

Lesser Cost: FPGAs offer more cost-effective solutions and best choice for the prototyping of the ASICs.

The development of the algorithms using the AI/ML based platform is one of the difficult tasks. Here we can make it easier by using the high level synthesis.

For more details on the AI/ML based FPGA and hardware accelerators refer the manuscripts on the AI/ML FPGA Design.

13.6 Important Highlights

Following are few of the important points to summarize this chapter.

1. Use the low power aware design techniques and pipelined logic to design the processor.
2. The clocks and resets should be fully controllable.
3. Use the DFT friendly designs to improve the controllability and observability of the design.
4. The reduction in the voltage levels to have low power enabled mode can be one of the best strategy in the standby or low power mode.
5. Do not use the latches in the design, if latches are used then they should be controlled by the clocks.
6. Using the gray encoding the power can be optimized.
7. Using the one-hot encoding method the timing can be improved but it increases the area.

Bibliography

1. www.springer.com : https://link.springer.com/book/10.1007/978-981-19-4652-3
2. www.springer.com : https://doi.org/10.1007/978-981-16-5955-3

V. Taraate, *Advanced Digital Design Techniques*,
https://doi.org/10.1007/978-981-95-4510-0

Index

0–9
2:1 mux, 20
2:4 decoder, 76
2:4 decoder having active low outputs, 77
2-input XOR, 17
2-varibale K-map, 99, 104
32-bit pipelined processor logic, 193
4-bit ring counter, 107
4:1 mux, 69
4:16 decoder, 79
4:2 encoder, 81
8-bit arithmetic unit, 58

A
Active high asynchronous reset, 144
Active low asynchronous reset, 144, 160, 167
AI accelerators, 203
ALU, 63
Always on logic, 179
Analog to Digital Converter (ADC), 1
AND, 11
AND arrays, 84
AND gate, 11
AND logic, 39
Area constraints, 4, 17, 33
Area optimized design, 24
Arithmetic unit, 56
Arrival time, 118
Asynchronous designs, 119
Asynchronous FIFO, 145
Asynchronous reset, 143

B
Balanced scan chain, 188
Binary Coded Decimal (BCD), 3
Binary encoding, 153
Boolean equations, 107, 110
Bus interfaces, 181
Bypass, 200
Bypass multiplexers, 188

C
Caches, 199
Cascade multiplexers, 69
CDC techniques, 181
CLK mux, 28
Clock and data path, 200
Clock and reset, 194
Clock and reset management, 199
Clock buffers, 120
Clock domain crossing, 123
Clock domains, 121
Clock gating, 174, 175, 199
Clock gating cell, 177
Clock gating test, 190
Clock latency, 120
Clock mux, 28, 124, 126
Clock muxing, 126
Clock mux logic, 50
Clock polarity, 120
Clock skew, 120, 129
Clock to Q delay, 113
CMOS NOT gate, 5
Combinational logic, 3
Complement logic, 39
Config and status registers, 181
Controllability, 200

V. Taraate, *Advanced Digital Design Techniques*,
https://doi.org/10.1007/978-981-95-4510-0

Control path, 37
Control path synchronizers, 123
CPF is Common Power Format, 179
Custom ASIC gates, 45, 46
Cycle consumption, 193

D
Data and control path, 123
Data path, 30, 35
Data path selector, 48
Data path synchronizers, 124, 145
Data selectors, 54, 69
Debug logic, 182
Debug mux, 182
Decode, 196
Decoding logic, 76
Delay optimization, 15
Depth, 124
DFT friendly architecture, 199
DFT friendly design, 4, 31
DFT techniques, 186
Digital to Analog Converter (DAC), 1
Divide by 4 synchronous binary up counters, 98
Dynamic power, 173, 175
Dynamic voltage and frequency scaling, 178
Dynamic Voltage Drop (DVD), 175

E
Edge sensitive, 113
Efficient ALU design, 65
Encoder, 81
End point, 131
Excitation input, 97, 101, 106, 109
Excitation table, 96, 99, 101, 104, 156, 160, 164, 168
Execution, 196

F
Fast arriving signal, 70
Fetch, 196
FIFO, 124, 145
FIFO depth calculation, 124
FPGA LUT, 84
Frequency gating, 178
FSM-based controllers, 172
FSM design, 158, 167
Functional blocks, 181

G
Gated clocks, 199
Glitches, 30, 75, 126, 176
Glitch free clock gating, 177
Glitch free clock mux, 126
Glitch free FSMs, 172
Gray codes, 139
Gray counter, 102
Gray encoding, 154, 172
Grouping, 85

H
Handshake synchronizers, 141
Hazards, 71
Hex numbers, 3
Hold time, 113
Hold violation, 119

I
Input to reg path, 114
Insertion delays, 114
Instruction RAM, 193
Internally generated clock, 200
IO devices, 76
IO interfaces, 194
IR drop, 175
Isolation cells, 179, 180
Is ss shown, 158

J
Johnson counter, 108
JTAG, 182

K
K-map, 160, 168

L
Latches, 91
Late arriving signal, 70
Leakage current, 173
Level shifters, 179
Level synchronizer, 123, 127, 139
Logic duplication, 84
Logic gates, 9, 37
Logic unit, 60
Lower technology nodes, 199
Low power, 199
Low power aware architectures, 173
Low power cells, 7, 174
Low power design, 173

M
Maximum area, 5
Maximum frequency, 130
Mealy FSM, 151, 158, 161, 167, 169
Memories, 76
Memory controller, 196
Memory map, 87
Metastability, 121, 123
Metastable state, 123
Micro-architecture, 194, 197
Minimum area, 5
Mixed signal designs, 202
MOD 4, 98
MOD-4 synchronous binary down-counter, 102
MOD-4 synchronous binary up-counter, 99
Moore FSM, 151, 157, 165
Multi-bit data, 124
Multiple voltage islands, 178
Multiplexer, 19
Multiplexer 2:1, 37
Multi-Vdd, 178
Multivoltage power supply, 180
Mux, 19, 37
Mux synchronizer, 141

N
N:1 mux, 48
NAND, 12, 15, 17, 20
NAND gate, 12
NAND logic, 40
n-bit adder, 54
Negative clock skew, 129, 135
Negative setup slack, 119
Next state logic, 155, 162
NOR, 10, 15
NOR gate, 10
NOR logic, 42
NOT, 9
NOT gate, 9
Number systems, 2

O
One-cold encoding, 86, 87
One-hot encoding, 154, 172
Opcode, 55
Optimized design, 26
OR, 10
OR gate, 10
OR logic, 41
Output logic, 155, 162

P
Parallelism, 199
Parity checker, 85
Path delays, 71
Performance, 4
Pipelined, 137
Pipelined controller design, 193
Pipelining, 137
PLL, 181
PLL and clock circuits, 181
Polarity control, 120
Positive clock skew, 129, 134
Positive edge sensitive D flip-flop, 93
Positive setup slack, 118
Power, 6, 26, 154, 173
Power aware scan chain, 188
Power domains, 180
Power gates, 179
Power gating, 177
Power management unit, 179, 181, 199
Power optimization, 172
Power switch and testing, 190, 191
Power switches, 179, 180
Priority encoding logic, 82
Priority logic, 69
Processor, 1
Propagation delay, 8, 14, 50, 70, 86
Pulse synchronizer, 143

R
RAM, 87
Register balancing, 135
Reg to output path, 115
Reg to reg path, 118, 129
Required time, 118
Reset path, 97
Reset synchronizers, 143
Reset test logic, 201
Resource optimization, 55
Resource sharing, 26, 28, 53
Resource utilization, 4
Retention cells, 179
Retention registers, 180
Ring counter, 105
RISC processor, 194
ROM, 87

S
Scan chain, 186, 188
Scan flip-flop, 185
Scan mux, 31
Scan or the debug mode, 31

Scratch pad memory, 181
SDRAM controller, 197
Sequential design, 3, 6, 96
Setup + hold, 113
Setup slack, 118
Setup time, 113
Setup violations, 114, 119
Shift register, 120
Speed, 5
Standby modes, 199
Start point, 131
State encoding, 153
State register, 155, 162
State table, 98, 101, 104, 106, 109, 156, 160, 163, 168
Static-0 hazard, 72, 74
Static-1 hazard, 72
Store, 196
Supply nets, 179
Synchronization, 196
Synchronizers, 121, 199
Synchronous binary down counter, 100
Synchronous gray counter, 104

T
Testability, 200
Test mode, 187
The clock and reset management unit, 181
The Intellectual Property (IP), 179
The synchronous sequential circuit, 116
Timing, 92, 93, 95
Timing and area constraints, 5
Timing optimization, 135
Timing path, 118
Timing violation, 114, 137
Twisted ring counter, 111
Two-flop level synchronizer, 127

U
Universal logic, 19, 33, 37
Universal logic gate, 15–17, 45
UPF is Unified Power Format, 175, 179
UPF or CPF, 199

X
XNOR, 13
XNOR gate, 13
XNOR logic, 44
XOR, 12
XOR gates, 13, 85
XOR logic, 43

MIX
Papier aus verantwortungsvollen Quellen
Paper from responsible sources
FSC® C105338

If you have any concerns about our products,
you can contact us on
ProductSafety@springernature.com

In case Publisher is established outside the EU,
the EU authorized representative is:
Springer Nature Customer Service Center GmbH
Europaplatz 3, 69115 Heidelberg, Germany

Printed by Libri Plureos GmbH
in Hamburg, Germany